Emancipatory Climate Actions

"Why does the world seem ever further from averting climate catastrophe? Past decades have seen no shortage of high-profile rhetorics. For an international treaty commitment, public and private efforts are unprecedented. Viable technical and wider social solutions abound. Yet real progress remains thwarted. In this timely and innovative book, Laurence Delina points to some key reasons why—and offers some novel and important new insights. Extending beyond the usual policy tropes, the book bursts with fresh ideas for challenging deep hegemonies and collectively 'culturing' the needed radical transformations. The result is an invigorating triumph of hope over despair."

—Andrew Stirling, *Professor of Science & Technology Policy, SPRU—Science Policy Research Unit, University of Sussex, UK*

Laurence L. Delina

Emancipatory Climate Actions

Strategies from histories

Laurence L. Delina
Boston University
Boston, MA, USA

ISBN 978-3-030-17371-5   ISBN 978-3-030-17372-2   (eBook)
https://doi.org/10.1007/978-3-030-17372-2

This Palgrave Pivot imprint is published by the registered company Springer Nature
Switzerland AG
The registered company address is: Gewerbestrasse 11, 6330 Cham, Switzerland

Hope has two daughters. Their names are Anger and Courage.
Anger at the way things are, and Courage to see that they do
not remain the way they are.
—Attributed to *Augustine* of Hippo

*To Kuya Mat, Ate Kay, Ate Yang, Kuya Gab, and Insyang—and
million other Filipino children.*

PREFACE

Emancipatory Climate Actions offers strategies to strengthen the mobilization of and for contemporary climate actions. This book connects not only the fragmented pockets of opportunities for keeping fossil fuels in the ground and for accelerating a just transition to largely sustainable energy systems—essential in addressing the rapidly accelerating global climate challenge—but also envisages and advances a new hegemonic agenda for culturing contemporary human societies: a hegemony characterized by just emancipations and sustainable transformations. This book suggests some strategies mined from select historical episodes of four large-scale social movements in history. This book extends the conversation on how to strengthen climate actions—as a transformative and emancipatory force—that I have earlier argued in a 2018 Palgrave Pivot title called *Climate Actions: Transformative Mechanisms for Social Mobilisation*.

In this book, I envisage and design climate actions following lessons from four specific large-scale social movements in history. This book picks up on some of the key strategies imprinted in select mobilizations for: an independent India, highlighting the role of Gandhi, his salt making in Dandi, and the ensuing nonviolent protests in Dharasana; the modern American civil rights movement, underlining Rosa Parks and her arrest, and the ensuing bus boycotts in Montgomery; the building up of the anti-Marcos movement in 1980s Philippines, focusing on the grassroots mobilization following an assassination of Ninoy Aquino as he returned from exile, the rise of his widow, Cory, in the Filipino imagination of a democratic Philippines, and its culmination in the generally peaceful People Power Revolution of 1986; and the student-led Burmese pro-democracy movement in 1988, following the murder of a student activist in Rangoon, the expanding dissent across Burma, and highlighting Aung San Suu Kyi's failure to link necessary multilevel actions that could have changed the course of Burma's history.

The book is intended to benefit at least two sets of audiences. Its first target readers are climate activists, who use traditional tactics, such as warm-bodied protests, and those who are prefiguring desirable futures, such as those engaging in community renewable energy. The book provides these two groups of climate activists some strategies that they can consider and include in their own existing campaign repertoires. The book's second target readers are scholars working on areas of climate change, social justice, progressive actions, sustainability transitions, nonviolent conflicts, and social mobilizations.

Key in my evolving understanding of climate actions is my numerous encounters with many people from many places. Despite the limitations of my Philippine passport, I had the unique privilege of traveling the world during the last ten years, engaging with people—and learning along the way. This book saw first light in Sydney, Australia where I wrote my Ph.D. thesis under the mentorship of Mark Diesendorf. Mark was the first to suggest to study the grassroots dimension of climate actions and envisage strategies that might strengthen the Movement's ongoing work. I am very grateful that my path crossed with Mark. In 2013, I, Mark, and John Merson published an article in *Carbon Management* on what we can learn from histories of select social movements (see Delina et al. 2014, vol. 5, pp. 397–409). This book expands on that article.

I also had the opportunity to test some of my early ideas at Harvard, where, in 2013 and 2016, I joined Sheila Jasanoff's Science, Technology and Society (STS) "laboratory," and at the Rachel Carson Center in Munich, where, in 2017, I was a Rachel Carson Fellow. I have encountered not only bright minds in these places but also made friends that I continue to nurture up to these days. The Frederick S. Pardee Center for the Study of the Longer-Range Future at Boston University, my academic home since 2015, offers me a productive space to pursue my research interests and become a productive scholar. At the Pardee Center, I thank Anthony Janetos, our Center director and my postdoc supervisor, for his unwavering support; Cynthia Barakatt for patiently reading and commenting on my work; Theresa White and John Prandato for their assistance. I am very grateful that I was able to immerse myself in these vibrant communities.

Despite having spent most of my time living outside the Philippines for almost eleven years now, I still call the Philippines home. Filipinos are not historically responsible for the current climate change but we are at the forefront of the changes it brings to our physical and social environments. The vulnerability of my home country to extreme climate impacts always make me think about my family, especially my nephews and nieces—who are still young children as of this writing—knowing that their future is almost locked in to be a future of suffering. The histories of social mobilizations for the common good, however, make me remain hopeful that the future of my nephews and nieces will be better. Urgent climate actions are necessary not only to effectively and successfully bend the curve of rising greenhouse gas emissions but also, and more importantly, to advance new cultures of a just, emancipatory, and transformative world. I sincerely hope that that kind of future awaits my nephews and nieces and many other young Filipinos to whom I dedicate this book.

Boston, MA, USA   Laurence L. Delina
January 2019

CONTENTS

1 Introduction: Emancipatory and Transformative
  Climate Actions   1

2 Four Histories of Social Mobilizations: Dandi,
  Dharasana, Montgomery, Manila, and Rangoon   17

3 Visioning and Identity-Building: An Overarching
  Vision for Heterogeneous Campaigns   35

4 Culturing and Framing: Working on the Ills
  of the Past, in the Present, for Tomorrow's Benefits   53

5 Triggering Communal Peer Pressure: Spreading
  a Shared Understanding of Demands   71

6 Boosting Publicity: Old and New Media, Deliberations,
  and Organic Ideology Articulation   83

7 Diversifying Networks: Webbing Heterogeneous
  Actors and Their Plural Campaigns   97

8   Conclusion: Strengthening Climate Actions Through
    Emancipatory and Transformative Mobilizations   111

Index   117

CHAPTER 1

Introduction: Emancipatory and Transformative Climate Actions

Abstract Climate change, which evidences the impacts of humanity's fossil fuel-based choices, is breaking down social, political, and economic order, with people living in poorer places unjustly carrying more of its burdens. Despite an international agreement to reduce present and future emissions, a number of governments all over the world continue their support on fossil fuels. With myopic climate actions from these governments, the time is ripe for strengthening the grassroots dimension of climate actions: to scale up efforts from citizen-oriented protests against the fossil fuel regime and to expand instances of prefiguring desirable futures. Beyond these, the climate action movement also needs to elevate the discourse about a counter-hegemony that would replace the neoliberal capital order, which, for generations, props up the fossil fuel-based paradigm of progress. Doing so requires the Movement to embark on expansive campaigns that advance alternative politics, new economics, and desirable social cultures.

Keywords Climate change · Neoliberalism · Hegemony · Emancipatory transformation · Just sustainability · Climate mobilization · Climate action · Social movement

Unabated, the global climate system continues to breakdown, signaling an ominous social, political, and economic present and future for human societies. Sustainable energy technologies to replace fossil fuel-based

© The Author(s) 2019   1
L. L. Delina, *Emancipatory Climate Actions*,
https://doi.org/10.1007/978-3-030-17372-2_1

systems largely responsible for climate change are now making possible perpetual and environmentally sustainable forms of power production from wind, water and sunlight, especially as their costs decline. But we still feel incapable of evading or controlling the forces and powers that would address our unquenching thirst for fossil-fueled development and facilitate a rapid, large-scale transition to planetary sustainability. Our political, economic, social, and cultural systems are strongly embedded within the fossil fuel regime, and, thus, a complex web entwines the abstractions of modern life with the materiality of our technological systems and infrastructures.

We are now in an era where the consequences of our neoliberal, fossil fuel-based, capitalist choices manifest not only in extreme modifications in weather events costing lives and livelihoods mostly in poorer communities in developing countries (where people with the least contribution to climate change also live) but also in electoral democracies gone miserably awry; territorial borders functioning as gates to hell; jobless neighborhoods turning violent; informal and poor urban settlements rising in densities; and intellectualism, science, and higher education subjected to persistent attacks.

Our present world seems to be leading us into a corrosive, dystopic future world of inequality, hatred, neocolonialism, racism, misogyny, paranoia, authoritarianism, and unsustainability. Genuine transformative change now seems beyond the capacity of our present technological, political, and economic systems. We can no longer rely upon capitalism's self-correction mechanisms. With crises—natural, physical, social, economic, and political—gathering steam, politics and economies wilt. This ongoing paralysis urgently requires a new vision of the future—a new ideology, hegemony, and intellectual leadership—spurred by a large-scale public clamor for transformative and emancipatory change that could be delivered through simultaneous climate actions. This book offers one way forward to achieve this vision.

For many countries, especially in high emission countries of Australia and the United States, power structures at the national level remain in a state of political gridlock on climate change response, resulting in ineffective policymaking and action (Wishart 2019; Washington 2018; Dunlap and McCright 2011; Van Rensburg and Head 2017). Despite the widely hailed Paris Agreement on climate change, myopia in government response to the climate challenge in these countries—and beyond—is evidenced by the low ambition in the Nationally Determined Contributions that make up the Agreement; commitments which, even when collectively met, will not ensure a future that is climate-safe for everyone.

Australia and the United States continue to use and support coal. The United States has, under a climate denying President, hosted side events in the last two meetings of Parties to the United Nations Framework Convention on Climate Change touting coal. Other high emission countries are also at fault. In China—where coal plant construction stopped in some provinces (not because of climate considerations but due to overcapacities), coal continues to account for about two-thirds of electricity generation, and additional plants have been permitted. China also exports its coal combustion technologies through its Belt Road Initiative. Japan, Indonesia, and Turkey continue to proceed with their coal plant construction plans (Climate Action Tracker 2018). Beyond coal-addicted governments, mainstream media, still an inarguably key source of information for many people, are also being remiss in reporting the current state of climate and the necessary climate actions (Miliauskas and Anderson 2016; Bacon and Nash 2012; cf. Yacoumis 2017).

Fossil fuels continue to be taken out of the ground and transported to points of consumption, where they are burned to fill societies' seemingly insatiable demand for energy. The neoliberal capitalist orientation of many countries in the world—where climate actions are primarily seen as countering "development" objectives—lends a hand to climate action gridlock. Interestingly, even those in the fossil fuel regime had acknowledged, as early as 1982 in an Exxon-supported symposium on climate change at Columbia University, the flaws of free markets when it came to climate change (Rich 2018).

On the first weekend of October 2018 in Incheon, South Korea, the International Panel on Climate Change (IPCC 2018), an institution tasked by governments to advance the science of climate change, released a special report confirming a sobering picture that has long been painted by climate scientists of the potentially disastrous impacts of allowing global mean surface temperature to rise by an additional 1.5 °C compared with pre-industrial levels. The report details more extreme weather events, sea level rise, and ocean acidification affecting crops, wildlife, water availability, and human health. But the report remains a relatively conservative assessment of the consequences of climate change as it leaves out key details as to the damaging impacts to specific populations, who will be displaced first and forced to migrate hence increasing chances of conflicts. The report also failed to discuss the fat tails of climate change impacts, the tipping point s in the climate system, which, when manifest, could lead to irreversibility and acceleration of change. The cost of doing the necessary climate actions, such as through pricing carbon, is also limitedly discussed.

The 1.5 °C report, nonetheless, points out to the absolute necessity of doing effective and sufficient climate actions *now*. These ambitious actions require urgency and unprecedented changes, which according to the IPCC, are still affordable and feasible. The question, however, persists: with world leaders disregarding the required systemic change— that is, to quickly transform not only the global energy infrastructure but also the culture of consumerism, which is strongly hinged in the neoliberal capitalist agendas of many high emitting states, both historical and current—where could we fathom a new vision that would lead to real transformative change in human society? With time running out, how can we mobilize climate actions and create a significant dent in the present mood of complacency, hopelessness, fear, and despair? In other words, how can we advance transformative and emancipatory climate actions now?

In liberal states, power structures are supposedly dependent upon people's consent and therefore can be revoked at any time. However, it seems that political power in many democratic states at present does not necessarily come from the will of the majority. The rise of populist leaders, including in my own country, the Philippines, was paved by minority votes. Rodrigo Duterte was elected with a plurality of just 39% of the overall vote (Salaverria 2016). In the United States, where I work, Donald Trump won just under 46% of the popular vote (*The New York Times* 2016).

Interestingly, populist leaders such as Duterte and Trump are also leaders who do not pay attention to climate actions. Duterte, in his first State of the Nation Address, less than two months of his assumption into office, signaled his pro-coal stance (Rappler 2016). Trump, a climate denier himself, is pulling out the United States out of the Paris Agreement and is watering down efforts to address climate change from almost everywhere he can: repealing Barack Obama's Climate Action Plan (Smith 2017; Friedman and Plumer 2017); proposing financial guarantees for coal and nuclear power plants (Crooks 2017); reviving the Keystone XL oil pipeline (Smith and Kassam 2017); opening part of the Arctic National Wildlife Refuge in northern Alaska (King 2018) and almost all offshore waters for oil and gas drilling (Milman 2018a);

announcing plans to halt car fuel efficiency standards (Davenport 2018); and proposing cuts in the Environmental Protection Agency's budget (Tabuchi 2017) and the axing of $2 billion funding for the Green Climate Fund and 20% of the IPCC's funding (Office of Management and Budget 2018).

Of course, the United States is not alone in the effective and sufficient climate action impasse (although some states have become outliers, such as California). Other high emission nation-states are also busy not with instituting climate actions but with curtailing freedoms of their citizens: Russia poisons government dissenters (Urban 2018; Harding 2016); Saudi Arabia dismembers its critics (McKernan 2018). With states not working to address the urgent and scaled need for climate actions now while becoming unaccountable, extensive public pressure needs to be exerted.

There are historic precedents for successfully exerting pressure on intolerant, hostile, antagonistic, paranoid, and selfish powerholders to resolve issues of important public interest. These public exercises of power have also led to large-scale social changes. Histories show that social movements could indeed cause powerholders to be toppled, and new cultural and social norms to be created—and sustained. Achieving large-scale, socially transformative and emancipatory ends—these histories tell—require tremendous, strong, and systematic mobilization (Sharp 1973; Ganz 2010; Moyer et al. 2001).

The high stakes of climate inaction and the slow, inadequate, and ineffective response from many governments would require a similarly scaled citizen-oriented response for public mobilization—a role that, for some time now, has been attached to the climate action movement. The climate action movement comprises individuals, groups, networks, alliances, and coalitions, offering multiple and heterogeneous activities, campaigns, strategies, and tactics to realize change across multiple levels and spaces. Activating and strengthening the "people power" dimension of climate actions comes at a time when the Movement has already seen successes and experienced failures to reflect upon. This book summons select moments in the histories of some large-scale social mobilizations to gather strategies that the climate action movement may learn from and can incorporate in their repertoires. Additionally, social movement scholars find that mobilizations could influence media attention, agenda setting, government funding, and legislative success (e.g. Agnone 2007; Amenta et al. 2010; Olzak et al. 2016).

Climate actions, particularly through outward-oriented, nonviolent protests, have indeed made progress when demonstrators stopped the Keystone XL pipeline from being constructed (Goldenberg and Roberts 2015), when kayakers blocked Shell's drilling rigs in the Seattle harbor (Hackman 2015), when endowments and pension funds, begun divesting from fossil fuel holdings (Carrington 2018; Milman 2018b), when local governments in developing countries such as the Provincial Governments of Bohol and South Cotabato in the Philippines deny coal-based development (Conde 2018; Sarmiento 2018), etc. Meanwhile, community energy—the principal modus operandi of the German *Energiewende* (Morris and Jungjohann 2016)—has expanded in Europe (Bauwens et al. 2016; Mey and Diesendorf 2018; van der Schoor et al. 2016; Oteman et al. 2014), Australia (Hill and Connelly 2018; Hall et al. 2010), Central America (Madriz-Vargas et al. 2018), and Asia, such as in Thailand (Delina 2018) and Indonesia (Thomas et al. 2018). In the light of our current political, economic, and social challenges and circumstances, mobilizing more climate actions using these tested and proven strategies—while fielding innovative ones—has become more important.

A public constituency waiting to be mobilized to strengthen climate actions exists. This constituency is found in high emission states where climate actions currently have low influence in policymaking and media reporting. A 2017 survey by the Yale Program on Climate Change Communication finds that 69% of registered voters in the United States endorse the Paris Agreement (only 13% oppose) and that 78% support regulations and taxes to address climate change (only 10% oppose) (Marlon et al. 2017). The same Yale Program reports, in 2018, that 85% of Americans support funding more research into renewable energy; 77% support regulating carbon dioxide as a pollutant; 70% support setting strict carbon limits on existing coal-fired power plants: and 68% support requiring fossil fuel companies to pay a carbon tax (Marlon et al. 2018). When there is a conflict between environmental protection and economic growth, 70% of their survey respondents think the former is more important than the latter (Ballew et al. 2018). A longitudinal study from Stanford University also finds that the majority of Americans support many climate-friendly energy policies, including renewable energy targets, limitations on emissions by utilities, and energy efficiency standards, and are even willing to pay some amount to have them enacted (Krosnick and MacInnis 2013). Even in the Republican state of Texas,

79% of survey respondents agree that emissions should be reduced from power plants, while 76% agree that their government should limit emissions from businesses (Krosnick 2013).

A similar proportion of population in Australia—another high emission state—agrees that climate action is necessary. The Commonwealth Scientific and Industrial Research Organisation (CSIRO) finds that four in five Australians (81%) think climate change is happening and increased investment in renewable energy and public transport should be made (Leviston et al. 2014). These numbers reveal a silent and sympathetic, yet dispersed and "underutilized," majority, who can be potentially mobilized into the climate action movement.

Mobilizing climate actions from public constituencies, who want to address climate change within their means, capacities, and strengths, is not a small task. Mobilization is essentially about organizing numbers to wield political power. This is inarguably herculean; but when successfully organized, the strength of a number of people, small groups, and organizations "united together" for a common goal can result in desirable and durable social change. Gene Sharp (1973: 7), who has studied many nonviolent social movements, acknowledges the necessity "to wield power in order to control the power of threatening political groups or regimes." Saul Alinsky (1971: 113), considered by some to be the father of modern community organizing, writes: "Change comes from power, and power comes from organization. In order to act, people must get together." The many histories of social movements, indeed, suggests that affecting change is only possible if citizens would organize their numbers and become a strong political power (Sharp 1973; Ganz 2010; Della Porta and Diani 2006; Diani 1992).

These histories are littered with stories of successes and failures signaling that it is not unusual for movements to feel that they are not winning at some point of their struggles. The heterogeneity of successful and failed campaigns is distributed unevenly on a longer-term continuum. In this continuum lies observable moments by which (1) the degrees of public perception vary—from mere notice to actual engagement and pouring out of support; and (2) the scales are tipped off—when old regimes are finally displaced and new ones dominate. The climate action movement is not exempted from these vagaries in the evolutions of social mobilizations.

Already, the climate action movement has picked up on the potentials of history-inspired campaign strategies. Divestment, a case in point, is a strategy based from the anti-Apartheid movement in South Africa, and is used in calling for non-support of fossil-based systems. Largely successful outcomes have been seen in colleges, universities, local governments, and pension funds. But other lessons from histories of social mobilizations are yet to be unpacked, critically analyzed, and strategically advanced for their inclusion in the campaign repertoires of contemporary climate actions. This book attempts to explore these "other" lessons.

The book distinctively contributes to strengthening the Movement by assisting not only in harnessing existing momentum and sustaining ongoing actions but also in envisaging further opportunities for scaling (i.e. getting everyone involved and engaged), while understanding that not everything can be scaled. Since the powerful fossil fuel regime complex is successful in sustaining myopia, division, and ineffectiveness of climate actions in governments, the Movement needs to ratchet up its strategies to target the core structure, the backbone, and the foundation of our collective challenge—yet, one that is least discussed and confronted: the neoliberal capitalist system by which our contemporary social, political, and economic infrastructures are hinged (cf. Malm 2016; Battistoni 2018).

This book approaches the challenge of toppling the hegemony of the neoliberal capitalist order by heeding some of the lessons from four select histories of transformative and emancipatory social mobilizations. The book then attempts to use these lessons to craft strategies that could strengthen climate actions. These select histories function as reserves of ideas, which are summoned as case studies to understand *how* the "mechanisms" for effective social actions were generated and to explain *why* events and people offer such mechanisms. Through an iterative process (i.e. looking at the four histories broadly and collectively), these mechanisms would justify and explain successes and failures. Analyzing not only successes (which are privileged in the literature since they are the ones which are commonly reported and analyzed) but also failures reduces potential selection bias. These mechanisms are then used as structural frames in in-depth comparisons between yesterday's social actions and today's climate action movement to advance some future strategies.

There are obviously hundreds of details that show how comparisons between eras and struggles are similar and different. This book, thus, pays attention not only on the production of parallel strategies that arise from analyzing points of similarities between the past and the present but also in highlighting some points of divergence. This approach addresses another potential criticism where "actual fields of action" are impossible to observe in a historical assessment for obvious reason. While time travel is indeed physically impossible, one can still argue that inferring from records of histories yield important data.

A study of the strategies pursued by past social mobilizations to inform contemporary strategies for strengthening the climate action movement is a big project in itself. The insights to be gained from this analysis, however, are vital, rendering the perceived burden almost irrelevant. The results would benefit not only scholars and students alike but also—and perhaps most importantly—those who seek this study's practical contributions: the mobilizers for climate actions.

To make this work manageable, nonetheless, this book makes an important qualification concerning case selection. Three reasons rationalize the choice of four histories under study. First, the cases should show asymmetrical conflicts between adversaries in terms of strength of resources and capacity. This rationale is important for the book's general objective where non-state climate action groups are pitted against resource superior opponents in the fossil fuel regime complex. Second, the cases should exhibit diversity of both causes and effects that would allow the formation of some generalizations that are broader and stronger compared to what can be derived from a single case. Third, the cases should contain variations in terms of campaign outcomes: successes and failures, to allow a comparison of the mechanisms—what caused the outcomes—which are the principal interests of this book.

Measuring successes and failures is a task that is difficult to gather and, most especially, to defend. As mentioned above, a successful or failed campaign does not necessarily universalize, conclude, or totalize mobilizations since Movements exist in a continuum. To simplify the process of judging successes and failures, the study applied a two-step test. Test 1 asks: Was there an observable effect on the outcome such that the outcome could be plausibly interpreted as a direct result of a "mechanism"? Test 2 asks: Were the campaign's immediate stated goals fully achieved? A simple affirmative response to both questions automatically describes success; a negative in either one means otherwise.

By layering the assessment in this two-level test, a reflective analysis is done both from external and internal points of view. The first level of assessment (Test 1) contemplates the Movement's engagement as observed from outside the Movement itself. The second level (Test 2), meanwhile, reflects the internal goals, objectives, and aims of several groups that constitute a particular social movement and the specific objectives of a respective campaign. This approach to describing success and failure in campaigning is empirically and theoretically grounded. Kelly (2002) and Miller (1998), for instance, suggest that assessments must not be confined to a single metric but must include both internal and external aspects of organizational functioning. The two-way test employed in the analysis in this study includes both.

Among the many large-scale social mobilizations available to study, this book had selected moments in the social action histories of four large-scale social movements based on the above criteria. Chapter 2 describes these select moments in more details. Summoning these select stories—pinning down the footprints they left in their respective dynamics—helped explain why social actions were successful or not. This process led to the determining of at least five mechanisms that strengthen social actions: (1) clarity and coherence of vision; (2) narratives, stories and symbols; (3) peer pressure; (4) expansion using traditional and innovative forms of communication; and (5) pluralism and reflexivity. I present each of these mechanisms, respectively, as chapters in this book. In a way, these mechanisms reflect similar mechanisms I found in my study of contemporary social action groups, which I covered in detail in my other Palgrave Pivot book entitled *Climate Actions* (Delina 2019)—albeit with different organization.

Chapter 3 discusses clarity and coherence of vision in building a new collective identity to transform people's behaviors and ways of life such that they will actively engage with social action. This chapter advances a new expansionist, yet transformative and emancipatory, agenda aimed at countering the neoliberal capitalist order, which has been providing power to the fossil fuel regime, which, in turn, is driving atmospheric greenhouse gases to their extreme volumes.

Chapter 4 on messaging advances the idea that "culturing" and framing are necessary considerations in addressing climate denial and dissonance. Culturing—that is, developing an idea or message with active consideration of the local context and norms of the intended audience—mediates meaning. Framing delineates what is relevant and what is not, focuses people's attention to an issue, and helps tie elements so that a context-specific storyline rather than another is told.

Chapter 5 on triggering communal peer pressure relies on a proven formula in catalyzing a sense of obligation amongst social groups. Because communal peer pressure varies from person to person and group to group, it is challenging to describe its processes. But something is common in this social phenomenon: the process of spreading peer pressure.

Chapter 6 on boosting publicity describes how people's attention can be nudged by traditional and innovative communication tools. The chapter describes the significant role of mass media—despite their less than sterling coverage of social actions—and expands on the need of creating trustworthy platforms in social media, especially given the proliferation of fake news in these sites. The chapter highlights a key role for deliberative democracy exercises and underlines the need to co-opt intellectual organizations to build the foundations of a counter-hegemony to the neoliberal capitalist order.

Chapter 7 on diversifying networks take heed of the plural and cosmopolitan nature of the required climate actions. As a constellation of heterogeneous actors and groups, the climate action movement needs to be webbed so that each actor's unique contribution is accounted for in the collective work. As an informal arrangement, without a hierarchy and a central authority, the Movement will thrive on interconnections; hence, is best envisaged as a polycentric system.

Unlike other books on climate actions, including my other Palgrave Pivot title, this book argues for an expansionist order that webs these key mechanisms—which do not occur in series but in indeterminable and surprising ways. Each of the four histories under study started out as localized campaigns but expanded into Movements of national scale because they were clear of their expansionist agendas. The necessity of expanded climate actions mirrors the same intention since these activities, exercises, and processes intend to address not only local and national issues, but a global challenge.

The contemporary climate action movement, after all, involves no particular race, political ideology or country but the whole of humanity regardless of identity: skin color, language, cultures, political inclination, or citizenship. The universality of climate actions further implies that this mobilization is also a fight not only for present humans but also for the yet-to-be-born generations of humans, non-human species, natural ecosystems, and other abstractions such as "the future." The intended results of climate activism, thus, extend far beyond changes in community ordering, political leadership, or legislation (e.g. stronger community energy arrangements, effective climate legislations, shifts toward green politics, and 100% renewable energy transition)—although, of course, these shifts are, by all means, essential measures of rather incremental successes. The Movement has to focus on nothing less than the larger fundamental rethink of the future of human civilizations: a shift toward a new, better, durable, desirable, just, and sustainable order. This entails advancing a counter-hegemony to the failures of the neoliberal capitalist system that triggered climate change and can be done by mobilizing a Movement that offers alternative politics, new economics, and desirable social cultures.

REFERENCES

Agnone, J. (2007). Amplifying public opinion: The policy impact of the U.S. Environmental Movement. *Social Forces, 84,* 1593–1617.

Alinsky, S. (1971). *Rules for Radicals: A Pragmatic Primer for Realistic Radicals.* New York: Random House.

Amenta, E., Caren, N., Chiarello, E., & Su, Y. (2010). The political consequences of social movements. *Annual Review of Sociology, 36,* 287–307.

Bacon, W., & Nash, C. (2012). Playing the media game: The relative (in)visibility of coal industry interests in media reporting of coal as a climate change issue in Australia. *Journalism Studies, 13,* 243–258.

Ballew, M., Marlon, J., Maibach, E., Gustafson, A., Goldberg, M., & Leiserowitz, A. (2018). *Americans are More Worried about Global Warming, and Show Signs of Losing Hope.* Yale University and George Mason University. New Haven, CT: Yale Program on Climate Change Communication.

Battistoni, A. (2018, August 3). How not to talk about climate change. *Jacobin.* https://bit.ly/2Dwn4Br.

Bauwens, T., Gotchev, B., & Holstenkamp, L. (2016). What drives the development of community energy in Europe? The case of wind power cooperatives. *Energy Research & Social Science, 13,* 136–147.

Carrington, D. (2018, September 10). Fossil fuel divestment funds rise to $6tn. *The Guardian.* https://bit.ly/2MhV8SV.

Climate Action Tracker. (2018, May 3). Paris tango. Climate action so far in 2018: Individual countries step forward, others backward, risking stranded coal assets. https://bit.ly/2yQkNhI.

Conde, M. (2018, May 27). Bohol's no-coal ordinance 'an important victory'—Advocate. *Rappler.* https://bit.ly/2IRBf8y.

Crooks, E. (2017, October 1). US delivers electric shock with coal and nuclear subsidy plan. *The Financial Times.* https://on.ft.com/2Pci24H.

Davenport, C. (2018, August 2). Trump administration reveals its plan to relax car pollution rules. *The New York Times.* https://nyti.ms/2AAyIf9.

Delina, L. L. (2018). Energy democracy in a continuum: Remaking public engagement on energy transitions in Thailand. *Energy Research & Social Science, 42*, 53–60.

Delina, L. L. (2019). *Climate Actions: Transformative Mechanisms for Social Mobilisation*. Cham: Palgrave Macmillan.

Della Porta, D., & Diani, M. (2006). *Social Movements: An Introduction* (2nd ed.). Malden: Blackwell.

Diani, M. (1992). The concept of social movement. *The Sociological Review, 40*, 1–25.

Dunlap, R., & McCright, A. (2011). Organized climate change denial. In J. Dryzek & R. Schlosberg (Eds.), *The Oxford Handbook of Climate Change and Society* (pp. 144–160). Oxford: Oxford University Press.

Friedman, L., & Plumer, B. (2017, October 9). EPA announces repeal of major Obama-era carbon emissions rule. *The New York Times*. https://nyti.ms/2g5m8va.

Ganz, M. (2010). *Why David Sometimes Wins: Leadership, Organization, and Strategy in the California Farm Worker Movement*. Oxford: Oxford University Press.

Goldenberg, S., & Roberts, D. (2015, November 6). Obama rejects Keystone XL pipeline and hails US as a leader on climate change. *The Guardian*. https://bit.ly/2SNAYVg.

Hackman, R. (2015, May 16). Seattle 'kayak-tivists' take in Shell in battle over Arctic oil drilling. *The Guardian*. https://bit.ly/2F3EBUu.

Hall, N., Taplin, R., & Goldstein, W. (2010). Empowerment of individuals and realization of community agency: Applying action research to climate change responses in Australia. *Action Research, 8*, 71–91.

Harding, L. (2016). *A Very Expensive Poison: The Definitive Story of the Murder of Litvinenko and Russia's War with the West*. London: Guardian Faber Publishing.

Hill, D., & Connelly, S. (2018). Community energies: Exploring the socio-political spatiality of energy transitions through the Clean Energy for Eternity campaign in New South Wales. *Australia, Energy Research & Social Science, 36*, 138–145.

IPCC. (2018). Summary for Policymakers. In V. Masson-Delmotte, P. Zhai, H.-O. Pörtner, D. Roberts, J. Skea, P. R. Shukla, A. Pirani, & W. Moufouma-Okia, C. Péan, R. Pidcock, S. Connors, J. B. R. Matthews, Y. Chen, X. Zhou, M. I. Gomis, E. Lonnoy, T. Maycock, M. Tignor, & T. Waterfield (Eds.), *Global Warming of 1.5°C. An IPCC Special Report on the impacts of global warming of 1.5°C above pre-industrial levels and related global greenhouse gas emission pathways, in the context of strengthening the global response to the threat of climate change, sustainable development, and efforts to eradicate poverty* (p. 32). Geneva, Switzerland: World Meteorological Organization.

Kelly, L. (2002). *Research and Advocacy for Policy Change: Measuring Progress*. Brisbane: The Foundation for Development Cooperation.

King, P. (2018, April 19). Trump administration takes first steps toward drilling in Alaska's Arctic refuge. *Science.* https://doi.org/10.1126/science.aat9366.

Krosnick, J. (2013). *Public Opinion on Global Warming in Texas: 2013.* California: Woods Institute for the Environment, Stanford University.

Krosnick, J. A., & MacInnis, B. (2013). Does the American public support legislation to reduce greenhouse gas emissions? *Daedalus, the Journal of the American Academy of Arts & Sciences, 142,* 26–39.

Leviston, Z., Price, J., Malkin, S., & McCrea, R. (2014). *Fourth Annual Survey of Australian Attitudes to Climate Change: Interim Report.* Canberra: CSIRO.

Madriz-Vargas, R., Bruce, A., & Watt, M. (2018). The future of community renewable energy for electricity access in rural Central America. *Energy Research & Social Science, 35,* 118–131.

Malm, A. (2016). *Fossil Capital: The Rise of Steam Power and the Roots of Global Warming.* London: Verso.

Marlon, J., Fine, E., & Leiserowitz, A. (2017, May 8). Majorities of Americans in every state support participation in the Paris agreement. *Yale Program on Climate Change Communication.* https://bit.ly/2qSMAeO.

Marlon, J., Howe, P., Mildenberger, M., Leiserowitz, A., & Wang, X. (2018, August 7). Yale climate opinion maps 2018. *Yale Program on Climate Change Communication.* https://bit.ly/2M8Ml9Y.

McKernan, B. (2018, October 31). Jamal Khashoggi strangled as soon as he entered consulate, prosecutor confirms. *The Guardian.* https://bit.ly/2OhBZS2.

Mey, F., & Diesendorf, M. (2018). Who owns an energy transition? Strategic action fields and community wind energy in Denmark. *Energy Research & Social Science, 35,* 108–117.

Miliauskas, A., & Anderson, H. (2016). Uncertainty framing and the IPCC 'Fifth assessment report'. *Australian Journalism Review, 38,* 143–154.

Miller, A. (1998). *Strategic Management* (3rd ed.). Boston: Irwin McGraw-Hill.

Milman, O. (2018a, January 4). Trump administration plans to allow oil and gas drilling off nearly all US coast. *The Guardian.* https://bit.ly/2QjMbvt.

Milman, O. (2018b, January 10). New York City plans to divest $5bn from fossil fuels and sue oil companies. *The Guardian.* https://bit.ly/2EuYXk9.

Morris, C., & Jungjohann, A. (2016). *Energy Democracy: Germany's Energiewende to Renewables.* Cham: Palgrave Macmillan.

Moyer, B., McAllister, J., Finley, M. L., & Soifer, S. (2001). *Doing Democracy: The MAP Model for Organizing Social Movements.* Gabriola Island: New Society Publishers.

Office of Management and Budget. (2018). *America First: A Budget Blueprint to Make America Great Again.* Executive Office of the President of the United States. https://bit.ly/2nvjrBO.

Olzak, S., Soule, S. A., Coddou, M., & Muñoz, J. (2016). Friends or foes: How social movement allies affect the passage of legislation in the U.S. Congress. *Mobilization, 21,* 213–230.

Oteman, M., Wiering, M., & Helderman, J.-K. (2014). The institutional space of community initiatives for renewable energy: A comparative case study of the Netherlands, Germany and Denmark. *Energy, Sustainability and Society, 4,* 11.

Rappler. (2016, July 25). Full text: President Duterte's 1st State of the Nation Address. *Rappler.* https://bit.ly/2ANAdVz.

Rich, N. (2018, August 1). Losing earth: The decade we almost stopped climate change. *The New York Times Magazine.* https://nyti.ms/2WfsSXy.

Salaverria, L. B. (2016, May 11). Duterte passes Aquino in number of votes won. *The Philippine Daily Inquirer.* https://bit.ly/2yUwST7.

Sarmiento, B. S. (2018, July 3). South Cotabato board rejects coal mining. *The Philippine Daily Inquirer.* https://bit.ly/2RCL5et.

Sharp, G. (1973). *The Politics of Nonviolent Action.* Boston: Porter Sargent.

Smith, D. (2017, March 28). Trump moves to dismantle Obama's climate legacy with executive order. *The Guardian.* https://bit.ly/2nez5Al.

Smith, D., & Kassam, A. (2017, January 24). Trump orders revival of Keystone XL and Dakota access pipelines. *The Guardian.* https://bit.ly/2jtvMUy.

Tabuchi, H. (2017, April 10). What's at stake in Trump's proposed EPA cuts. *The New York Times.* https://nyti.ms/2nyyZrW.

The New York Times. (2016, August 9). Presidential election results: Donald J. Trump wins. *The New York Times.* https://nyti.ms/2PcFpLz.

Thomas, S., Richter, M., Lestari, W., Prabawaningtyas, S., Anggoro, Y., & Kuntoadji, I. (2018). Transdisciplinary research methods in community energy development and governance in Indonesia: Insights for sustainability science. *Energy Research & Social Science, 45,* 184–194.

Urban, M. (2018). *The Skripal Files: The Life and Near Death of a Russian Spy.* New York: Henry Holt and Company.

Van der Schoor, T., van Lente, H., Scholtens, B., & Peine, A. (2016). Challenging obduracy: How local communities transform the energy system. *Energy Research & Social Science, 13,* 94–105.

Van Rensburg, W., & Head, B. W. (2017). Climate change skeptical frames: The case of seven Australian sceptics. *Australian Journal of Politics & History, 63,* 112–128.

Washington, H. (2018). Denial—The key barrier to solving climate change. *Encyclopedia of the Anthropocene, 2,* 493–499.

Wishart, R. (2019). Class capacities and climate politics: Coal and conflict in the United States energy policy-planning network. *Energy Research & Social Science, 48,* 151–165.

Yacoumis, P. (2017). Making progress? Reproducing hegemony through discourses of "sustainable development" in the Australian news media. *Environmental Communication, 12,* 840–853.

CHAPTER 2

Four Histories of Social Mobilizations: Dandi, Dharasana, Montgomery, Manila, and Rangoon

Abstract History is rich of accounts of nonviolent social movements toppling down powerful, yet unjust regimes. This chapter reviews four select moments of some of these histories: Gandhi's salt march (which triggered the Free India Movement); Rosa Park's arrest (which triggered the modern American civil rights movement); Ninoy Aquino's assassination (which triggered the Philippine People Power Revolution ousting a dictator); and the murder of Phune Maw (which triggered the Burmese Uprising). This chapter describes, in some key details, these localized tipping moments, how subsequent macro-mobilizations had transpired, and the results of the ensuing campaigns.

Keywords Gandhi · March to Dandi · Dharasana campaign · British India · Rosa Parks · Martin Luther King, Jr. · Montgomery · Bus boycott · Civil Rights Movement · Ninoy Aquino · Cory Aquino · People Power Revolution · Burmese Uprising · Aung San Suu Kyi

This chapter reviews four histories of social mobilization, focusing on select episodes that can offer ideas and strategies for contemporary application. The historical episodes described in this chapter include: (1) the 1930 events in the Indian Movement for Self-Rule, looking particularly at Gandhi's March to Dandi between 12 March and 5 April where he gathered salt and the Dharasana salt mine campaign in May; (2) the 1955–1956 bus boycotts that catalyzed the modern African-American

L. L. Delina, *Emancipatory Climate Actions*,
https://doi.org/10.1007/978-3-030-17372-2_2

Civil Rights Movement, focusing especially at the arrest of Rosa Parks on 1 December 1955 and the following 13-month bus boycott in Montgomery, Alabama; (3) the anti-Marcos movement culminating in the 1986 Philippine People Power Revolution, which was spurred by the mobilization activities by the Philippine Catholic Church and progressive entities in urban and rural communities; and (4) the 1988–1990 Burmese Uprising, triggered by the murder of a university student in Rangoon and spontaneously expanded into a nationwide anti-junta and pro-democracy campaigns. The first three moments facilitated successful campaigns, while the fourth one failed to accomplish its intended goal.

1930 INDIA: GANDHI'S MARCH TO DANDI, THE DHARASANA DISPERSAL, AND THE MOVEMENT FOR FREE INDIA

The year 1930 is considered the pivotal year for the 90-year-long Indian Movement for Self-Rule (Weber 1997). That year, two episodes ignited a massive social shake-up and challenged the dominion of the British Empire in the Indian subcontinent. With Mahatma Gandhi, the independence campaign started to adopt nonviolent resistance including the Dandi March from 12 March to 5 April, where Gandhi defied the British salt tax laws, and the Dharasana campaign in May, which ended in violent beatings of peaceful campaigners. These two episodes attracted international attention, spurring the international community to question the British imperial dominance over the region. The Movement successfully concluded seventeen years later on 14 August 1947 with the passage of the *Indian Independence Act of 1947* at the British Parliament. India finally became an independent nation but lost Pakistan, which became a separate country.

The Movement began 90 years earlier, in 1857, when British-controlled India saw its first mutiny against British rule. This rebellion and the ensuing mutinies came to be known as India's First War of Independence. More radical, often militant, campaigns, and agitations for independent India continued in the early part of the twentieth century, but these were all effectively suppressed. It was from 1930 onwards, when the Gandhian nonviolent discipline and resistance were adopted, that the Movement finally catapulted to success. Of the many events that led to the expulsion of British colonialism in the subcontinent, Gandhi's Dandi March turned the tide. Pictures of Gandhi gathering salt in a non-violent protest against the British salt monopoly became the iconic image representing the Movement worldwide (Weber 1997).

Gandhi and his followers walked 400 kilometers over 26 days from Sabarmati Ashram near present-day Ahmedabad to the seaside village of Dandi in the state of Gujarat where he started making salt for himself. Marching was a familiar campaign tactic for Gandhi, who had led a five-day march under harassment from authorities in South Africa. Gandhi arrived in Dandi on the morning of 5 April, a day ahead of schedule. At dawn the next day, Gandhi stood on the shore, bent down, and picked up a clump of mud. This simple, yet symbolic, gesture of salt-making sent the signal to the tens of thousands on the beach, who, following Gandhi's example, started collecting saltwater in buckets and pots and boiled it to produce salt (Singhal 2010). Hundreds of thousand others on beaches along India's 6500-kilometer-long coastlines systematically followed this civil disobedience (Dalton 1993: 115).

With millions breaking the British salt laws, the Dandi campaign provided a visceral symbol that strengthened the Movement and turned it into a larger, subcontinental Movement. The episode helped trigger the spread of mass civil disobedience across India. A wave of resignations by village officials followed, with almost a third of all village officials in Surat district resigned by the first week of April (Brown 1989: 104–105). The British Government, in retaliation, arrested more than sixty thousand people by end of March 1930 (Dalton 1993).

But the Movement did not back down. In the following weeks, the salt campaign expanded to other forms of nonviolent protests, including boycotts of British cloth and other goods. The civil disobedience of 1930, for the first time, involved Indian women, who became active participants in many other campaigns (Basu 1995; Thapar 1993). Women carried jugs to the shore to boil water for salt, boycotted auctions of confiscated goods, were beaten by police, and went to jail. In Bombay City, the salt satyagraha and cloth boycotts were widespread and marches were well-organized. In the Midnapur region of Bengal, peasants turned villages into forts after a police attack on salt-making volunteers (Sarkar 1987: 94–95). In Peshawar, Abdul Ghaffar Khan's followers picketed liquor stores (Sarkar 1989: 288).

Two months after Gandhi's salt-making at Dandi, he organized another major peaceful protest. Still following strict nonviolent discipline, he planned a raid of the Dharasana salt mines in May 1930 to goad the British authorities to arrest him. Gandhi wrote the Viceroy of India, Lord Irwin, of this intention and was duly arrested on 4 May even before the Dharasana campaign could start (Dalton 1993: 112). Despite Gandhi's arrest and the threat of violence, the raid went ahead as planned. As protesters marched from Dandi to Dharasana, several of the organizers, including Gandhi's wife, Kasturbhai, were arrested. Many other peaceful protesters were violently dispersed along the way. As the remaining protesters, led by Indian poet Sarojini Naidu, tried to encroach Dharasana's salt pens on 21 May, police begun charging and clubbing them. This violent dispersal became the second defining moment for the Movement, triggering an international call for Britain to finally give up India. At the end of 1930, *Time Magazine* named Gandhi its Man of the Year, comparing his salt march with New Englanders' defiance of the British tax on tea.

By January 1931, the British Government, through the Viceroy of India released Gandhi from prison. Lord Irwin and Gandhi signed an agreement that resulted in the removal of the salt tax. While considered a victory, some scholars argue that it was in fact a capitulation on Gandhi's part and the Movement more broadly. With the events of 1930 not resulting in immediate independence, Ackerman and Kruegler (1994: 194, 200), for instance, contend that "the campaign was a failure" and "a British victory" because Gandhi "gave away the store." Jawaharlal Nehru also viewed the 1930–1931 campaigns as lost opportunities.

Viewed in the short term, there is logic in this argument. However, for momentum-driven Movements such as a national independence from foreign powers with a long-term goal of shifting large-scale pub- lic opinion on the issue, the events of 1930—from the Dandi March to the Dharasana campaign to other nonviolent protests—were effec- tive catalysts for the Movement for two important reasons. First, these episodes won more popular support, hence lending to the "mobiliza- tion" aspect of the Movement. Second, these episodes built the capac- ity of the Movement by strengthening its base, hence underlining its resilience.

The year 1930 was, indeed, important for the Movement because it demonstrated that ordinary Indians had the power to drive events that they could organize and mobilize across scales and spaces. In the long run, the events of 1930 led to the British realization that they could no longer take for granted the reliability of Indians who staffed their bureaucracies. It also led to an international recognition of the legitimacy of the Movement and the weakening British grip on the subcontinent.

Although India did not become an independent nation until seventeen years later, the trigger events at Dandi, Dharasana, and all across India, served as the first major cracks that helped in the disintegration of the British colonial establishment. Louis Fischer's (1954 [1982]: 102) widely-read biography of Gandhi remarked of the salt march: "Technically, legally, nothing had changed. India was still a British colony...[and yet, after the salt march], it was inevitable that Britain should some day refuse to rule India and that India should some day refuse to be ruled." The suffering of the dissenters did not change British minds but it did change the minds of Indians about the British.

1955 United States: Rosa Parks' Arrest, the Montgomery Bus Boycott, and the American Civil Rights Movement

The year 1955 can be viewed as the pivotal year for the modern Civil Rights Movement in the United States, a little more than one year after the US Supreme Court issued a ruling on *Brown v. Board of Education* that declared the unconstitutionality of state laws establishing separate public schools for white and colored students. African-Americans sensed that change was finally coming, but *Brown* alone was not sufficient to bring the cause for equal rights into national focus. On 1 December 1955, Rosa Parks was arrested for simply claiming a seat on a bus, triggering a thirteen-month bus boycott in Montgomery, Alabama. This episode ignited the national campaign for civil rights of the modern era. The wide range of effective nonviolent activities following Parks' arrest and the subsequent bus boycott led to the achievement nine years later of important civil rights legislations that ended racial segregation and enforced equal voting rights in the United States.

At the moment Rosa Parks was arrested, no one on that Montgomery bus knew that a spark had just ignited that would power the Movement forward, eventually changing the course of history of racial equality. On the night of her arrest, Parks remarked: "My resistance to being mistreated on the buses and anywhere else was just a regular thing with me and not just that day" (Morris 1984: 51). Parks' small refusal to give up her seat, however, was a key historical episode in itself. That event spawned a series of largely nonviolent mass actions that tilted the battle over civil rights from a campaign fought in state legislative houses and courts into a large-scale national social action movement.

Parks was arrested for violating Alabama's bus segregation law, which stated that colored people had to occupy only the back area of a public bus. Parks (1992: 113–117) recalled that event in her own words:

> When I got off from work that evening of December 1, I...[took] the Cleveland Avenue bus home...I saw a vacant seat in the middle section of the bus and took it...The next stop was the Empire Theater, and some whites got on. They filled up the white seats, and one man was left standing. The driver looked back and noticed the man standing, then he looked back at us. He said, "Let me have those front seats," because they were the front seats of the black section. Didn't anybody move. We just sat right where we were, the four of us. Then he spoke a second time: "Y'all better make it light on yourselves and let me have those seats."...I could not see how standing up was going to "make it light" for me...The driver... saw me still sitting there, and he asked was I going to stand up. I said, "No." He said, "Well, I'm going to have you arrested." Then I said, "You may do that."...He got out of the bus and stayed outside for a few minutes, waiting for the police...Eventually two policemen came. They got on the bus, and one of them asked me why I didn't stand up. I asked him, "Why do you all push us around?" He said to me, and I quote him exactly, "I don't know, but the law is the law and you're under arrest."

In January 1956, one month after Park's arrest, Montgomery's black population would rise up and boycott city buses, ending the boycott only when the law segregating public transport was finally lifted. The boycott crippled the bus line, drew tens of thousands of protesters to nonviolent demonstrations and mass actions, introduced the young Martin Luther King, Jr. to America as one of the nation's charismatic voices, and sparked a nonviolent Movement that eventually spread outside Montgomery, Alabama to towns in other southern States.

Parks' arrest, it turned out, was a strategy already planned for some time. Earlier on, the Montgomery branch of the National Association for the Advancement of Colored People (NAACP), where Parks was the secretary, believed that a high-profile court case to challenge bus segregation laws was necessary to publicize the cause. The local association waited for some time for this appropriate case to arrive and had been preparing campaign strategies when it occurred. When Parks was arrested, the branch knew what to do: stage a large-scale bus boycott involving all African-Americans in the town that would be sustained until segregation was lifted (Hohle 2013).

The boycott proved to be exceptionally effective, especially when religious ministers rallied their communities. Instead of taking public buses, carpooling was organized with many African-American car owners volunteering their cars and even white housewives driving their African-American servants to work. Others cycled and walked. This form of direct action went on for more than one year. The loss of many riders from Montgomery's transport system caused economic distress to the city. Pressure increased and went beyond Alabama's borders as other cities duplicated the bus boycott.

A little more than six months since Parks' arrest, a federal court, on 4 June 1956, ruled on the unconstitutionality of Alabama's racial segregation laws for buses. The State of Alabama appealed the decision but on 13 November 1956, the US Supreme Court upheld the lower court's ruling. On 20 December 1956, nineteen days after the first anniversary of Parks' arrest, all seats on the buses at Alabama were finally made available to anyone regardless of passenger's skin color (Hohle 2013; Parks 1992; Kohl 2000).

Following the Montgomery bus boycott, a series of major nonviolent campaigns came to characterize the Civil Rights Movement. Ensuing influential nonviolent civil disobedience and protests included: the 1960 Greensboro sit-ins in North Carolina, which led to the Woolworth department store chain removing its racial segregation policy in the southern United States; the 1961 Freedom Rides, where activists would ride interstate buses into the unsegregated southern United States to challenge the non-enforcement of US Supreme Court rulings on their unconstitutionality; the 1963 Birmingham campaign, led by Martin Luther King, Jr. to bring attention to the integration efforts of African-Americans in Birmingham, Alabama; the 1963 March on Washington, where Dr. King delivered his historic "I Have a Dream" speech calling for an end to racism; and the 1964 Freedom Summer campaign in Mississippi, a volunteer campaign to attempt to register as many African-American voters as possible in the State.

By the end of 1965, two national laws had passed, fulfilling the two important objectives of the American Civil Rights Movement. On 2 July 1964, after weeks of filibustering from Southern Senators, the US Congress passed the *Civil Rights Act of 1964*, which was then signed into law by President Lyndon B. Johnson. The Act put an end to racial discrimination and segregation based on race, color, religion, sex, or national origin in employment practices and public places. One year later, on 6 August 1965, Congress passed the *Voting Rights Act of 1965* that prohibits discrimination in voting on account of race and color, suspending poll taxes, and recognizing the rights of illiterate people to vote. Today, African-Americans from the South vote in large numbers, and occupy seats in their state legislatures and in the US Congress. In 2009, an African-American president was elected.

1983–1986 PHILIPPINES: NINOY AQUINO'S ASSASSINATION AND THE PEOPLE POWER REVOLUTION

On 21 August 1983, 20,000 Filipinos waited outside Manila International Airport. In the VIP lounge, the 73-year old mother of Benigno "Ninoy" Aquino, his siblings, in-laws and close friends were waiting; also in the vicinity were more than a thousand-armed soldiers and police. Aquino had been the leading figure of the opposition before Ferdinand Marcos imposed martial law in 1972, and was imprisoned for seven years before being allowed medical care abroad. That day, Aquino was going home from his exile in Boston. When the jet carrying him stopped at the gate, soldiers emerged from an awaiting patrol van. Three of them boarded the plane, spotted Aquino, and escorted him to the door. After stepping out, one policeman blocked the door, stopping a TV cameraman and news reporters. Aquino was then hustled through a side door down a stairway to the tarmac. Within seconds, a gunshot was heard, followed by a volley of shots. When other passengers emerged, Aquino was seen on the pavement, mortally wounded. Nearby lay another body later identified as that of Ronaldo Galman, who was accused of the murder.

The Marcos era in Philippine politics arrived as the country transitioned from a series of colonial takeovers. Ferdinand Marcos ruled from 1965 and stayed in power for the next twenty-one years. Marcos perpetuated his power by systematically removing constitutional prohibitions on the executive branch through a martial law declaration in 1972. *Proclamation 1081*—the executive order paving the way for martial law—allowed Marcos to consolidate his executive power by arresting opposition leaders, shutting down media outlets, centralizing and monopolizing state institutions and services, restructuring the armed forces, and institutionalizing nepotism and cronyism in government (Nepstad 2011; Wurfel 1977; Overholt 1986; Boudreau 2004;

McCoy 1999). During his incumbency, Marcos issued decrees nullifying constitutional limits to power, eliminating the office of the vice president, shutting down or taking over media establishments, suspending habeas corpus, and arresting political opponents—turning himself into a dictator (Thompson 1995: 52, 54).

Marcos and his wife Imelda, whom he appointed governor of Metro Manila and later a cabinet minister, locked up political control. Insisting on a cut of every major business transaction in the Philippines, the duo amassed great wealth through their cronies in state monopolies, bilateral and multilateral aid, and loans from international financial institutions (Aquino 1984; Overholt 1986; Wurfel 1977, 1988; Hutchcroft 1991). Debt from the International Monetary Fund and the World Bank expanded during martial law from $2.7 billion in 1972 to $10.7 billion in 1977 (Thompson 1995: 65–66). Dissenters, meanwhile, languished in prison; with more than sixty thousand political arrests made in five years. With the opposition either in prison, in exile, or collaborating with the Marcos regime, opposition was essentially absent (Schock 2005; Wurfel 1977). The Communist Party of the Philippines (CPP), which was declared a state enemy, went underground and began expanding its armed component, the New People's Army (NPA), geographically both in the countryside and in urban centers (Villegas 1986; Overholt 1986).

Without a strong visible political opposition, the Philippine Catholic Church became the last remaining social institution to serve as a channel for political opposition (Youngblood 1978). The Church worked to bring together the various strands of the opposition by mobilizing various grassroots groups, which eventually led to their convergence in the early 1980s (Schock 2005; Youngblood 1978). Cadres from the Church and leftist organizations persistently organized the grassroots and raised consciousness among the urban poor, farmers, fisher-folks, students, and the middle class (Zunes 1999). A network of sectoral organizations, concerned with diverse social action issues such as poverty, gender equality, labor, and education, was consequently formed (Schock 2005; Villegas 1986).

In 1980, a number of labor unions left the government-controlled organization called the Trade Union Congress of the Philippines to form the progressive *Kilusang Mayo Uno* (KMU; translation: First of May Movement). Also that year, non-Marcos-allied capitalists formed the Makati Business Club (MBC). Under mounting domestic pressure, Marcos lifted martial law in 1981. Despite this, the number of extra-judicial killings continued to increase and new restrictions banning mass protests were enacted (Boudreau 2004; Wurfel 1977).

In 1983, the Philippines faced a major economic downturn. The Marcos Government was running huge deficits to finance the bankrupt companies of his cronies (David 1996; Overholt 1986). At this point, Filipino business groups started to openly criticize the regime, with MBC loudly denouncing the growing public debt and state corruption (Overholt 1986). The Church, along with grassroots and labor groups, became even more critical of the regime's human rights abuses, corruption, and abuse of power. With business groups and the Church shifting into an active regime opposition, Ninoy Aquino was prompted to return to the country (Overholt 1986).

Ninoy Aquino's assassination was reported across the world but not in the Philippines. Government-controlled television and radio stations carried no news about it. Only *Radio Veritas*, the Roman Catholic radio station, reported the murder. Mourners filed through the living room of the Aquino's home in Manila to pay their respects to Ninoy whose body was laid in his bloodstained clothes. Ninoy's funeral started with the Manila archbishop, Jaime Cardinal Sin, saying a mass for him and eulogizing Aquino as having fallen in the cause of democracy. After the service, thousands walked through Epifanio de los Santos Avenue (EDSA), the widest and the longest thoroughfare in Manila, with another two million more waiting for hours to glimpse the procession (Overholt 1986). None of Ninoy's eleven-hour funeral-turned-protest was carried on local television; only *Radio Veritas* covered the event.

The conspiracy to kill Ninoy Aquino backfired; instead of eliminating a potential threat, a martyr was created. After the funeral, grassroots mobilization intensified and various strands of the political opposition begun converging (Schock 2005; Overholt 1986). On 9 September 1983, the day when the traditional Filipino mourning period for Ninoy's death ended, close to thirty thousand people turned out to support the first explicit anti-Marcos protest (Rocamora 1983). Two weeks later, at the eleventh anniversary of Martial Law on 21 September and thirty days after Aquino's assassination, close to one million people gathered at *Liwasang Bonifacio* (Bonifacio Park) in Manila (Aquino 1984; Rocamora 1983). Aquino's assassination mobilized the Catholic Church, the MBC, and liberal, moderate, and leftist politicians to form a unified Movement.

These new waves of mobilization, most obvious in urban areas, involved the middle class, professionals, and businesspeople.

MBC organized weekly anti-Marcos demonstrations with nearly a hundred thousand office workers in the country's financial district of Makati marching down the streets as supporters threw yellow pieces of shredded telephone directories from surrounding skyscrapers (Nepstad 2011; Zunes 1999; Diokno 1988; Tiglao 1988; Aquino 1984). In two years since Aquino's assassination, more than 150 nonviolent demonstrations took place in Makati and elsewhere (Thompson 1995).

With the widespread opposition increasing its political visibility, grassroots groups also expanded. Student organizations, labor unions, human rights groups, and groups of urban poor participated in rallies, joint sit-ins, and demonstrations and other forms of non-cooperation including successful general strikes where shops were closed and public and private transportation were stopped (Schock 2005; Zunes 1999). In 1985, the *Bagong Alyansang Makabayan* (BAYAN; translation: New Nationalist Alliance) was created as an alliance of progressive groups to coordinate mass actions. At the end of 1985, Marcos' legitimacy was severely weakened. Confident that the opposition could still be divided, Marcos called for "snap" or early elections for February 1986 (David 1996).

Traditional political parties, aware of the potential of achieving greater strength with greater numbers, began initiating various alliances. The United Democratic Opposition (UNIDO) was formed to bring in various political parties under one party umbrella. UNIDO also brought together women's groups, youth groups, nationalists, sectoral and regional forces, and civil libertarians (Rocamora 1983). The now united opposition, with the exception of the CPP and BAYAN who insisted on boycotting the elections, agreed to field Corazon "Cory" Aquino, Ninoy's widow, as its presidential candidate. During the campaign, Cory urged nonviolent discipline. Church leaders also insisted on nonviolence. An autonomous election-monitoring body, the National Movement for Free Elections (NAMFREL), which was tied to the Church, was formed to closely monitor the elections (Goldman and Pascual 1988). NAMFREL was able to mobilize and train more than two hundred thousand volunteers to conduct parallel counts in about 90 percent of the country's voting precincts (Goldman and Pascual 1988).

The 1986 elections, as expected, were heavily marked by widespread fraud. NAMFREL exposed serious discrepancies in the reports by the government-controlled elections commission who reported Marcos as the winner. In response, Cory, speaking before a crowd of nearly two million people, proclaimed victory for herself and announced a program of national nonviolent civil disobedience to protest the stolen elections (Zunes 1999; Aquino 1984). Before the campaign was even implemented, a different four-day large-scale social action mobilization took place at EDSA in what had become the "People Power Revolution"—a sobriquet reflecting the audacity and zeal of Filipinos.

The EDSA Revolution began with a military mutiny when Marcos's Defense Minister Juan Ponce Enrile and the Chief of Staff of the Philippine Armed Forces Fidel Ramos, along with a group of young military officers, planned an attack on the presidential residence. After their plans were discovered, the mutineers barricaded themselves in two major military camps along EDSA, announced their break from Marcos, exposed massive election cheating, and declared Aquino as the legitimate president. Later, Cardinal Sin made an appeal over *Radio Veritas*, asking the people to support the mutineers. In response, tens of thousands of ordinary Filipinos gathered along EDSA outside the rebel camps. Within a remarkable 77 hours, Filipinos from all walks of life and economic strata disobeyed government orders to evacuate the streets. Instead, a carnival-like atmosphere prevailed with the public singing, dancing, and holding prayer vigils (Astorga 2006; Arevalo 2000; David 1996). In the absence of support from government-controlled TV and radio stations, alternative media reported the ongoing revolution (Gonzales 1988).

Marcos deployed military tanks to EDSA in an attempt to disperse the growing crowd. A wall of tens of thousands of people, however, stopped this contingent and did not move even under threat. Instead, they sat down in front of the tanks and greeted the soldiers with flowers and prayers (David 1996). The tanks withdrew without a single shot being fired (Elwood 1997). After this retreat, a nationwide mutiny of soldiers ensued. Jet fighter pilots ordered to attack the rebel military camps refused to carry out their orders (Elwood 1997). As news of defections spread, people started celebrating. Crowds in EDSA continued to grow to hundreds of thousands. On the morning of 25 February, Corazon Aquino was sworn in as president (David 1996). Two hours later, Marcos took his own oath of office. His declared presidency, however, proved to be short-lived. With troops defecting in increasing numbers and with unwavering crowds at EDSA, he finally lost his grip on power and exiled to Hawai'i.

1988 BURMA: MURDER OF PHUNE MAW AND A FAILED UPRISING

On 12 March 1988, riot police shot and killed Phune Maw, a university student, in Burma's capital Rangoon. The killing occurred as students were protesting at the local police department over the release of a son of a Burma Socialist Programme Party (BSPP) officer who was implicated in an earlier brawl at the Sanda Win tea shop. (BSPP was the sole political party present in Burma at the time.) The killing angered pro-democracy groups and prompted students at the Rangoon Institute of Technology (RIT) to protest against police violence (Lintner 1990). The protests spontaneously expanded outside the capital and evolved from a police brutality issue into a nationwide anti-junta campaign. What had become the Pro-democracy Movement in Burma had some significant gains, including the temporary replacement of military with civilian rule in 1988, and a victory by the opposition party, the National League for Democracy (NLD) in multiparty elections in 1990. Still, the movement was not able to accomplish its intended goal to oust military rule and successfully transition to a democratically elected government.

Following decades of isolation, Burma, a year before the death of Phune Maw, faced major economic challenges (Taylor 1991; Guyot 1989). On November 1987, the government of Shu Maung, also known as Ne Win, withdrew its newly issued 25, 35, 75, and 100 kyat currency notes, leaving only 45 and 90 kyat notes in circulation. The decision was prompted by Ne Win's belief that the latter notes, being divisible by 9, were lucky. The decision promptly wiped out savings for tuition fees which angered many students, who ran riots through Rangoon (Guyot 1989). In December, the government issued a policy requiring farmers to sell their produce below market rates so that the government could collect more revenue. The policy sparked violent rural protests.

The murder of Phune Maw in March 1988 triggered a series of protests against Ne Win and his regime's economic and financial policies coupled with anger over continued regime repressions. When the killing went unpunished, large-scale student-organized protests ensued.

As students marched from Rangoon University toward RIT, five days after Phune Maw's murder, the police opened fire. In this so-called Bloody Friday massacre, about forty-one died according to Burmese officials, while other reports suggest that over two hundred were killed (Boudreau 2004; Fink 2001; Burma Watcher 1989; Silverstein 1990).

Government forces then occupied Rangoon University and arrested more than a thousand student protesters. Some students escaped and were joined by individuals from the working-class neighborhoods in several other demonstrations that spread throughout Rangoon. During these times, major Rangoon landmarks such as the Shwedagon Pagoda became rallying points of the uprising.

In response, the government promptly shut down schools and universities for several weeks. Upon the reopening of universities in late May 1988, students continued their campaign demanding the release of those who were earlier arrested (Schock 2005). The student-organized Movement disseminated leaflets and reorganized the strikes. By mid-June 1988, campaigns were taking place at all universities in the capital prompting the government to close Rangoon University. Despite this, large-scale nonviolent protests throughout the capital materialized with Buddhist monks, the urban poor, and factory workers joining thousands of student protesters (Schock 2005; Burma Watcher 1989). Their non-violent campaigns, however, were met with violence as riot police killed between eighty and a hundred unarmed protesters, while arresting hundreds of other activists (Schock 2005).

In response to the growing anti-junta campaign, the government held a special Party Congress in July where it was announced that the detained students would be released, and that Ne Win who led the junta since 1962, would resign. Sein Lwin, the military commander most responsible for the killings of the previous month, was announced as Ne Win's successor (Tucker 2001; Guyot 1989; Silverstein 1990). These political developments did not stop the protests.

In August, huge national protests were launched calling for the ouster of Sein Lwin and a Burmese democratic transition. The protests culminated with a general strike during an astrologically auspicious date, 8 August 1988 ("8888"). On that day, close to one million students, workers, monks, professionals, the unemployed, the urban poor, and members of various ethnic groups marched in the streets of Rangoon (Silverstein 1990; Mydans 1988). On the night of 8888, the *Tatmadaw* (the Burmese armed forces) opened fire on the crowd killing at least two thousand unarmed protesters (Sharp 2005; Schock 2005; Ghosh 2001). On 12 August, Sein Lwin suddenly resigned, an event that left many protesters confused but gleeful. A civilian, yet a prominent member of the ruling party, Maung Maung assumed the presidency later that day (Burma Watcher 1989; Guyot 1989).

The next day, tens of thousands denounced Maung Maung's accession to power and called for the end of one-party rule. In the next two weeks, hundreds of thousands of Burmese citizens from all classes, walks of life, and ethnic origins, participated in the nonviolent campaign, which included marches, demonstrations, rallies, and nationwide strikes (Burma Watcher 1989; Guyot 1989). During this period, Aung San Suu Kyi developed a strong following and became the public face of the Burmese opposition (Kaushikee 2012; Schock 2005; Suu Kyi 1995).

The 8888 general strike severely weakened the regime and many thought the junta would be dismantled (Schock 2005). By mid-September, however, opposition protests turned violent and unruly. Clashes between the *Tatmadaw* and the protesters became increasingly rampant. Opposition elites, including Aung San Suu Kyi, meanwhile, became divided to the extent that they could not even arrive at a common strategy to approach the junta (Schock 2005). The political elites also disregarded inputs from student leaders, failing to establish strong coalitions with them (Zin 2010).

On 18 September, amid the chaos and division among political opposition elites, a group of military generals, led by Saw Maung, successfully staged a coup and established a council promising to restore law and order and to prepare the country for a democratic election (Ferrara 2003; Silverstein 1990). Maung Maung's government was immediately dissolved and martial law was reimposed. The 8888 General Strike, which started a month earlier, collapsed as many Burmese workers, without food and money, returned to work.

By 21 September, the military regained control and the pro-democracy movement essentially crumpled. The government continued suppressing protests. Saw Maung's military council, however, scheduled elections for 1990, the first to be held in thirty years. The opposition parties mobilized for the elections and Aung San Suu Kyi, NLD leader, toured the country to solicit support (Silverstein 1990). On 27 May 1990, multiparty elections took place. Of the 447 parliamentary seats, Aung San Suu Kyi's NLD won 398, or about 80% of the seats, while the military council-backed party won only ten (Schock 2005). Instead of honoring their pledge to turn over power to the winning party, Saw Maung's military regime arrested and imprisoned opposition leaders, including Aung San Suu Kyi (Kaushikee 2012).

Unlike the Philippines case, the Burmese opposition was largely demobilized and too divided to seize the opportunity provided by the stolen 1990 elections through campaigns of noncooperation or civil disobedience (Zin 2010; Schock 2005; Fogarty 2008; Taylor 1991; Lintner 1990). As a result of this failure to maintain unity and cohesion of purpose, after more than two years of large-scale campaigning, during which Sein Lwin was ousted and multiparty elections were held, the Movement failed to achieve its ultimate aim to topple down a military regime and to transition Burma back to democracy.

REFERENCES

Ackerman, P., & Kruegler, C. (1994). *Strategic Nonviolent Conflict: The Dynamics of People Power in the Twentieth Century.* Westport: Praeger Publishers.

Aquino, B. A. (1984). Political violence in the Philippines: Aftermath of the Aquino assassination. *Southeast Asian Affairs,* 266–276.

Arevalo, C. G. (2000). Mary in Philippine catholic life. *Landas, 14,* 106–116.

Astorga, M. C. A. (2006). Culture, religion, and moral vision: A theological discourse on the Filipino people power revolution of 1986. *Theological Studies, 67,* 567–601.

Basu, A. (1995). Feminism and nationalism in India, 1917–1947. *Journal of Women's History, 7,* 95–107.

Boudreau, V. (2004). *Resisting Dictatorship: Repression and Protest in Southeast Asia.* New York: Cambridge University Press.

Brown, J. M. (1989). *Gandhi: Prisoner of Hope.* New Haven: Yale University Press.

Burma Watcher. (1989). Burma in 1988: There came a whirlwind. *Asian Survey, 29,* 174–180.

Dalton, D. (1993). *Mahatma Gandhi: Nonviolent Power in Action.* New York: Columbia University Press.

David, R. S. (1996). Re-democratization in the wake of the 1986 People Power Revolution: Errors and dilemmas. *Kasarinlan, 11,* 5–20.

Diokno, M. S. I. (1988). Unity and struggle. In A. Javate-de Dios, P. B. N. Daroy, & L. Kalaw-Tirol (Eds.), *Dictatorship and Revolution: Roots of People Power* (pp. 132–175). Manila: Conspectus.

Elwood, D. J. (1997). Philippines people power revolution, 1986. In R. S. Powers, W. B. Vogele, C. Kruegler, & R. M. McCarthy (Eds.), *Protest, Power, and Change: An Encyclopedia of Nonviolent Action from ACT-UP to Women's Suffrage* (pp. 412–414). New York: Garland Publishing.

Ferrara, F. (2003). Why regimes create disorder: Hobbes's dilemma during a Rangoon Summer. *The Journal of Conflict Resolution, 47,* 302–325.

Fink, C. (2001). *Living Silence: Burma Under Military Rule.* London: Zed Books.

Fischer, L. (1954 [1982]). *Gandhi: His Life and Message for the World*. New York: Mentor.

Fogarty, P. (2008, June 8). Was Burma's 1988 uprising worth it? *BBC News*.

Ghosh, A. (2001). Cultures of creativity: The centennial celebration of the Nobel Prizes. *The Kenyon Review, 23*, 158–165.

Goldman, R. M., & Pascual, H. A. (1988). NAMFREL: Spotlight for democracy. *World Affairs, 150*, 223–231.

Gonzales, H. (1988). Mass media and the spiral of silence: The Philippines from Marcos to Aquino. *Journal of Communication, 38*, 33–48.

Guyot, J. F. (1989). Burma in 1988: "Perestroika" with a military face. *Southeast Asian Affairs*, 107–133.

Hohle, R. (2013). *Black Citizenship and Authenticity in the Civil Rights Movement*. New York: Routledge.

Hutchcroft, P. D. (1991). Oligarchs and cronies in the Philippine state: The politics of patrimonial plunder. *World Politics, 43*, 414–450.

Kaushikee. (2012). Gandhian nonviolent action: A case study of Aung San Suu Kyi's struggle for Myanmar. *Gandhi Marg, 34*, 277–291.

Kohl, H. (2000). Rosa Parks and the Montgomery Bus Boycott. In J. Birnbaum & C. Taylor (Eds.), *Civil Rights Since 1787: A Reader on the Black Struggle* (pp. 443–456). New York: New York University Press.

Lintner, B. (1990). *Outrage: Burma's Struggle for Democracy*. Bangkok: White Lotus.

McCoy, A. W. (1999). *Closer Than Brothers: Manhood at the Philippine Military Academy*. New Haven: Yale University Press.

Morris, A. D. (1984). *The Origins of the Civil Rights Movement: Black Communities Organizing for Change*. New York: Free Press.

Mydans, S. (1988, August 12). Uprising in Burma: The old regime under siege. *The New York Times*.

Nepstad, S. E. (2011). *Nonviolent Revolutions: Civil Resistance in the Late 20th Century*. Oxford: Oxford University Press.

Overholt, W. H. (1986). The rise and fall of Ferdinand Marcos. *Asian Survey, 26*, 1137–1163.

Parks, R. (1992). *Rosa Parks: My Story*. New York: Dial Books.

Rocamora, J. (1983, November 16–18). *The Philippines after Aquino, after Marcos?* Paper from the 'Philippines after Marcos' conference, HC Coombs Lecture Theatre, Coombs Building, Australian National University.

Sarkar, T. (1987). *Bengal, 1928–1934: The Politics of Protest*. Delhi: Oxford University Press.

Sarkar, S. (1989). *Modern India, 1885–1947*. New York: St. Martin's Press.

Schock, K. (2005). *Unarmed Insurrections: People Power Movement in Nondemocracies*. Minneapolis: University of Minnesota Press.

Sharp, G. (2005). *Waging Nonviolent Struggle: 20th Century Practice and 21st Century Potential.* Boston, MA: Porter Sargent Publisher.

Silverstein, J. (1990). Civil war and rebellion in Burma. *Journal of Southeast Asian Studies, 21,* 114–134.

Singhal, A. (2010). The Mahatma's message: Gandhi's contributions to the art and science of communication. *China Media Research, 6,* 103–106.

Suu Kyi, A. S. (1995). Speech to a mass rally at Shwedagon Pagoda, August 26, 1988. In *Freedom from Fear and Other Writings.* New York: Penguin.

Taylor, R. H. (1991). Change in Burma: Political demands and military power. *Asian Affairs, 22,* 131–141.

Thapar, S. (1993). Women as activists: Women as symbols: A study of the Indian Nationalist Movement. *Feminist Review, 44*(Summer), 81–96.

Thompson, M. R. (1995). *The Anti-Marcos Struggle: Personalistic Rule and Democratic Transition in the Philippines.* New Haven: Yale University Press.

Tiglao, R. (1988). The consolidation of dictatorship. In A. Javate-de Dios, P. B. N. Daroy, & L. Kalaw-Tirol (Eds.), *Dictatorship and Revolution: Roots of People Power* (pp. 26–69). Manila: Conspectus.

Tucker, S. (2001). *Burma: The Curse of Independence.* Sterling: Pluto Press.

Villegas, B. M. (1986). The Philippines in 1985: Rolling with the political punches. *Asian Survey, 26,* 127–140.

Weber, T. (1997). *On the Salt March.* New Delhi: Harper Collins Publishers India.

Wurfel, D. (1977). Martial law in the Philippines: The methods of regime survival. *Pacific Affairs, 50,* 5–30.

Wurfel, D. (1988). *Filipino Politics: Development and Decay.* Ithaca: Cornell University Press.

Youngblood, R. L. (1978). Church opposition to Martial Law in the Philippines. *Asian Survey, 18,* 505–520.

Zin, M. (2010). Opposition movements in Burma: The question of relevancy. In S. L. Levenstein (Ed.), *Finding Dollars, Sense, and Legitimacy in Burma* (pp. 77–95). Washington, DC: Woodrow Wilson International Center for Scholars.

Zunes, S. (1999). The origins of people power in the Philippines. In S. Zunes, S. B. Asher, & L. L. Kurtz (Eds.), *Non-violent Social Movements: A Geographical Perspective* (pp. 129–157). Malden: Blackwell.

CHAPTER 3

Visioning and Identity-Building: An Overarching Vision for Heterogeneous Campaigns

Abstract For a mobilization to succeed in enrolling people to its cause, the public at large needs to be offered sound, clear, coherent and realistic descriptions of an alternative social order to the status quo in terms that they can readily understand and quickly relate to. In climate actions, visions of durable, desirable, just, emancipatory, sustainable and transformative futures can help build and culture new social identities, where members of the public will ably transform their prior understanding of their worlds and the ways they reflect this new understanding in their own present and future lives. To achieve this, visioning and identity-building in the climate action movement has to move beyond initiating, supporting and sustaining pockets of climate actions in small communities and people's centers of influence toward creating a larger imaginary that transforms the hegemony that orders contemporary human societies.

Keywords Vision · Identity · Civil disobedience · Nonviolence · Internet-based mobilization · Utopia · Cultural hegemony · Folk politics

Clear, coherent, and realistic visions are the first mechanism for social mobilization. These visions of the future offer social movements sound descriptions of an alternative order in terms that the public at large can readily understand and quickly relate to. Like master algorithms, clear visions of what the future ought to look like colors, solidifies, and constraints subsequent actions—the derivatives of such visioning exercises (Snow 2013; Moane 2011).

A strong vision also builds a strong identity. In the continuum describing social mobilizations, identity-building is the most important process (Diani 1992; cf. Lichterman 1999; Gamson 1991; Stryker and Burke 2000). During this phase, members of the public ably transform their prior understanding of the world and the ways they reflect this new understanding in their own lives. In transformations, people will start responding actively, decisively, and effectively to the social issue by not only involving themselves in political and social actions but also in changing their mindsets and behaviors (Moser 2009; e.g. Escobar and Alvarez 1992). In short, the public will prefigure what they want for themselves.

Identity-building transverses an "interactive and shared definition produced by several individuals (or groups at a more complex level)" (Melucci 1989: 34) and is "supposedly a *process* under continuous reflexive revision" (Saunders 2008: 230). When campaign participants are given a new sense of self-identity, historical mobilizations reveal that a Movement tends to strengthen. This new identity brings not only feelings but also, and most importantly, new experiences of solidarity, belongingness, harmony, and unanimity. Identity-building allows individuals, who may not be personally acquainted with one another but with which one, nevertheless, can share similar aspirations or visions of the future—a process that can lead to mobilization (Della Porta and Diani 2006; Melucci 1996; Hetherington 1998). Once strategies around identity-building are designed and deployed, individual actions—historical movements suggest—appear to become more self-directing (Lorenzoni et al. 2007; Moser 2009; cf. Stryker and Burke 2000). With self-direction, the chances of having successful campaigns are increased.

This chapter describes how the four social movements under study developed and deployed new visions of the futures they wanted within and across their respective heterogeneous constituencies, and how they adopted identity-building strategies to materialize and advance these visions. The chapter then describes how participants in the climate action movement have been laying down their visions of the future and how they build identities within their Movement. The chapter closes with a section on how visioning and identity-building could be embedded in the design of future climate actions, what these processes entail, and what they could look like.

VISIONING AND IDENTITY-BUILDING IN YESTERDAY'S SOCIAL MOVEMENTS

There are numerous and complex reasons why the histories under study sparked Movements that eventually expanded. But one critical factor, these histories suggest, is the *transformation* of large groups of citizens from being mere individual spectators into active participants and followers who became collectively organized, self-directing, and highly engaged social change groups. These histories also suggest that greater engagement was secured only after people realized three things: (1) that the Movement offered them a clear and unified alternative to their status quos; (2) that the Movement offered them a sense of community; and (3) that the Movement exemplified a set of moral values or virtues that they wanted to own for themselves.

Gandhi's vision of a free India, for instance, enjoined everybody, demonstrating how every Indian could and should contribute to the Movement. This almost-universality implied that ordinary Indians could participate in any campaign by targeting any area where the raj intruded into their lives and livelihoods. Gandhi demonstrated this through his civil disobedience performance in Dandi. At sea, Gandhi seamlessly connected his vision of a free India with a simple act of salt-making; yet, it had a bigger impact in terms of envisaging a free India not only from the British unjust salt monopoly but from its domination of the subcontinent. The Dandi March followed a well-thought-out, planned, and directed track that targeted a huge power base: India's subjugated masses and poor people (Weber 1997). The march tracked a path where recruitment potential would be high. But the larger objective of the march was to inculcate upon the Indian people a new sense of self, and the establishment of a new identity. This process of transformation toward a new Indian identity is best captured by a remark made by India's first Prime Minister, Jawaharlal Nehru:

> Of course these movements exercised tremendous pressure on the British Government and shook the government machinery. But the real importance, to my mind, lay in the effect they had on our own people, and especially the village masses. Poverty and a long period of autocratic rule...had thoroughly demoralised and degraded them. They had hardly any of the virtues necessary for citizenship...Non-cooperation dragged them out of the mire and gave them self-respect and self-reliance; they developed the habit of cooperative action; they acted courageously and did not submit so

easily to unjust oppression; their outlook widened and they began to think a little in terms of India as a whole... (Nehru 2005 [1958]: 154)

By effectively "transforming" the identities of the vast Indian collective during the salt march, Gandhi brought the Movement into its universal allure. In the following months, local organizers would identify issues around which the public could organize themselves along similar lines. Some examples include:

- In the fertile planes of Gujarat, north of Mumbai, where the Patidars live—the most unyielding practitioners of civil disobedience—the vital issue was land revenue. The British had viewed themselves as the ultimate landowners of the fields tilled by the Patidars. In an act of civil disobedience, the Patidars withheld their rent payments and, to prevent collectors from confiscating their belongings in lieu of the payments, packed up their valuables and cooking vessels and carted them across the state of Baroda (Hardiman 1981).
- In the Midnapur region, villagers refused to pay for watchmen resulting in a thousand resignation (Chakrabarty 1997: 104–107).
- In landlocked central provinces, masses of people—who were restricted access to forests and imposed high fees for grazing livestock—descended on forests, cut down bamboos, and grazed their cattle.
- In the urban centers of Bombay, Madras, Lahore, and Calcutta, campaigners seized control of public spaces where they would burn foreign cloths in bonfires (Masselos 1993: 71–83; Masani 1987: 91).

Movement-organized civil disobedience performances transformed the identities of the people who went through it. From a small group of elite, urban, and zealous Hindus wanting self-rule, the Movement expanded into a larger Indian public, which collectively imagine an independent India - one that overshadows religion, castes, and languages. The campaigns of 1930 resulted in an India becoming a new idea: a nation now sharing a common history. While down the line the subcontinent cracked into religious divisions, the events of 1930 galvanized Gandhi's Movement. The old order—British control, which rested comfortably on people's acquiescence and consent—was duly severed (Ackermann and Du Vall 2000: 109).

The four histories under study shared at least one important identity: **a high degree of nonviolent discipline**. Gandhi evoked its importance at all times. The African-American Civil Rights campaign in Montgomery also adopted similar nonviolent methods. Martin Luther King, Jr., a pastor, had found nonviolence in the Christian gospels in terms of the strong religious injunction of shunning violence and turned this aversion into the guiding principle of the Movement he soon dominated (Fredrickson 1995: 256–257).

In Montgomery, the principle of nonviolence manifested in African-Americans' communal transformation toward this new sense of collective identity. This transformation began as public commitment toward the bus boycott begun to wane. At this time, bombs were constantly thrown at many African-American residences and people began asking whether the boycott was indeed an effective campaign. When a bomb exploded in the yard of Martin Luther King, Jr. on 30 January 1956, the pastor immediately delivered a strong message that would define the new identity of the Movement from then on. He addressed the crowd:

> Now let's not become panicky. If you have weapons, take them home; if you do not have them, please do not seek to get them. We cannot solve this problem through retaliatory violence. We must meet violence with nonviolence...We must love our white brothers... no matter what they do to us. We must make them know that we love them. Jesus still cries out in words that echo across the centuries: 'Love your enemies; bless them that curse you; pray for them that despitefully use you.' This is what we must live by. We must meet hate with love. (King 2010 [1958]: 137–138)

The theme of King's message, one based on the doctrine of nonviolent discipline, was in many ways something King's listeners had not heard in this context before—a plea for overwhelming love and forgiveness of their attackers, along with a promise and a vision that it would bring them victory. The civil rights movement, for years before 1955, was primarily supported and sustained by the language of struggles and battles, of pains and sufferings, of retaliations and revenge. Following Gandhi and the examples of various campaigns such as those in Dandi and Dharasana, King's message of nonviolence gave African-American communities a new lens that eventually re-cast and transformed the bus boycott and the rest of the many other civil rights campaigns in a new and different light.

In the Philippines, calls for nonviolent discipline emanated from the Catholic Church, which during the 1986 elections campaign was instrumental in bringing in more support for the anti-Marcos movement (Zunes 1999). When an estimated 1.5 million Filipinos massed in Luneta Park in Manila the day after Marcos proclaimed his victory from a "snap" election of February 1986, Cory Aquino encouraged everyone to practice nonviolent civil disobedience. She called for a nationwide boycott of banks, newspapers, beverages, and movies. When the Movement culminated in the 1986 People Power Revolution, nonviolent discipline also became strongly evident as anti-Marcos protesters offered flowers and food to Marcos loyalist troops to appeal to their sense of nationalism (Schock 2005; Boudreau 2004).

The growing awareness of the power of nonviolent action allowed many Filipinos—now becoming more mindful of the need for social change but also anxious about the prospects of armed struggle pushed by the Communist Party of the Philippines and its armed wing, the New People's Army—to act in solidarity with the anti-Marcos Movement. The Catholic Church, which was particularly and strongly opposed to extrajudicial killings and other violent means of change, had been organizing and mobilizing based on strict nonviolent discipline, teaching people about the art of nonviolent civil resistance (Nepstad 2011). As a result of nonviolent anti-Marcos campaigns, many disaffected elites in the Philippine business community started switching sides in the early 1980s during which the Makati Business Club was formed (Zunes 1999). This coalition of businesses was key to securing financial support for the Movement (Overholt 1986; Aquino 1984). During the pivotal days after the stolen elections of February 1986, the Church urged, once again, Filipinos to use nonviolent action to protest. In a letter, the Church said:

> These [referring to the rigged election results] and many other irregularities point to a criminal use of power to thwart the sovereign will of the people…This means active resistance of evil by peaceful means…Our acting must always be according to the Gospel of Christ, that is, in a peaceful, nonviolent way. (Mercado 1986: 77–78)

One important qualification is in order when linking visioning and identity building. Most of Gandhi's followers took part in the Movement not solely so that they could seek moral transfiguration but to overcome their adversaries by denying them consent and revenues that made it possible for the British to continue their hold on India. The goal was not to demonstrate to the British that they were wrong but to force them out (Dalton 1993: 64). The goal of Aquino's nonviolent campaigns was also made along similar lines: those were not to demonstrate to the Marcos regime that they were wrong but to force the dictator out. This is important especially when traversing the terrain of social action campaigning where regimes try to buy in some time negotiating for some relief to delay, if not prevent, a transformation. When that situation arises, the histories of India and the Philippines are illustrative.

VISIONING AND IDENTITY-BUILDING IN TODAY'S CLIMATE ACTIONS

Salient in many present climate actions are confrontational exercises aimed at disrupting the technological systems attached to the fossil fuel complex—the regime which the climate action movement is attempting to dismantle (e.g. Bradshaw 2015; Dietz and Garrelts 2014; Sicotte and Joyce 2017). Through the years, the Movement has been clear on targeting this regime complex by steadily organizing and mobilizing supporters to protest the extraction of fossil fuel, including the mining and combustion of coal, and the building of new fossil fuel-based infrastructure such as coal-fired power-plants and oil and natural gas pipelines. Climate actions have also campaigned for divestments, urging schools, colleges, universities, churches, municipalities, cities, and pension funds, among others, to divest from their fossil fuel holdings (e.g. Fossil-Free 2018; McKibben 2013). Marches and rallies have also been deployed to raise public awareness on climate change and to gather support (e.g. Paliewicz 2018). The Movement also has recognized that protests are no longer produced and materialized exclusively in the streets and required warm human bodies (Pearce et al. 2015); Internet-based actions are now regarded as complementary, if not more important, modes of operation (Bennett and Segerberg 2012; Segerberg and Bennett 2011). Increasingly, recent climate action protests have been mobilized largely using social media (Wang et al. 2018; Hodges and Stocking 2015). Still, confrontational politics remain vital in the Movement's campaign repertoires.

Many climate actions also emphasize storylines that underline visions of climate-safe and sustainable futures. Many groups, for example, focus their mobilization efforts in highlighting alternative solutions to fossil fuel-based economies largely through examples of community energy (e.g. Aiken 2012). Community-based sustainable energy systems are often framed as visions of idyllic and romantic futures (e.g. Llewelyn et al. 2017), with many basing their storylines around the German *Energiewende* (Mey and Diesendorf 2018; Quitzow et al. 2016). Proponents of community energy would call on the German experience as an example of how small communities committing to energy transition in their localities could drive a national energy transition project.

Community energy offers a new vision for the Movement to rally on in that it can easily be championed. Community energy ranges from community-owned energy cooperatives to village-level energy efficiency programs to municipal-scale energy investments. These actions tend to be more personalized. Neighbors can participate and join together to exploit its many benefits (Van der Schoor and Scholtens 2015), in terms of energy autonomy (McKenna et al. 2016), energy autarky or self-sufficiency (Müller et al. 2011), and energy democracy (Delina 2018), among others. Community energy also tends to build identities among its participants as have been shown for example in an anecdotal evidence from my own work in Thailand (Delina 2018). Empirical data also show that community energy offers individuals and households new opportunities for income generation in terms of jobs and savings. Evidence further suggests that these economic benefits play far more important driving roles than environmental motivations for people to get involved with community energy (Seyfang et al. 2013; Yagatich et al. 2018; Bomberg and McEwen 2012).

But community energy alone cannot stand on its own. For climate actions to be effective, the energy transition has to be large enough to create a countervailing power to the fossil fuel regime complex. The Movement has no shortage of visions with regard to how that future can be achieved. Scholars have suggested what they call "the blooming of a thousand flowers" (Seyfang et al. 2013: 977; Foxon 2013: 19), where community energy systems would not only replace fossil fuels in villages and communities but large societies too. The Transition Towns movement is the oft-cited reference to this idea (Hopkins 2008; Aiken 2012; Connors and McDonald 2011); the transition projects of many cities, webbed together, is another (Bulkeley et al. 2010; Chatterton 2013).

At its core, the local transition storyline depicts a 100% renewable energy economy vision across local scales: from communities and villages to towns and cities, it has the potential for creating resilient and sustainable communities (Barr and Devine-Wright 2012). While initially focused on the local, this vision had already extended the use of renewable energy in the transition of states, nations, and even the entire planet (see list and descriptions in Delina 2016: 43–59). This vision suggests that renewable energy from wind, water, and sunlight could provide not only energy for electricity use but also for other sectoral uses such as public and active transport, industry, heating and cooling, aviation, and shipping. In this corpus, the work of Stanford academic, Mark Jacobson and his colleagues (2017), offer the oft-cited technical scenarios on how this vision could look like across 139 countries. And while the 100% renewable energy visions already permeate and began appearing in the campaign repertoires of many climate action groups, it is still loosely incorporated within the general goals of the Movement.

VISIONING AND IDENTITY-BUILDING IN TOMORROW'S CLIMATE ACTIONS

Envisaging a climate-safe future for everyone demands a utopian vision of a sustainable future *for all*—a vision beyond what can be captured in the 100% renewable energy vision and what can be communicated using storylines of community energy and confrontational politics. What is significantly missing in current climate action visioning programs is an encompassing agenda for a utopian world. This agenda should counter current dystopic images of hatred, inequality, and unsustainability, at the same time that this agenda also provides a mechanism that would move many in the climate action movement beyond their current focus on small-scale, community-based, local actions or what Srnicek and Williams (2016) termed as "folk politics."

Utopia concerns the "education of desire" (Levitas 2011), providing a frame, telling publics how and what to desire. Utopias would provide the Movement images of hope: that better futures are possible for everyone—not just the privileged few. Through this vision of better, robust, and desirable futures, utopian thinking helps spawn yearnings, generate actions, incite change, disrupt habits, and question consent to existing powerholders in the fossil fuel regime. Utopian thinking, thus, rejects melancholy and romanticism (that is somehow prevalent in folk politics) and instead invokes disappointment in the current state of affairs. Utopias, interestingly, were common tools for visioning in histories under study.

A focus on forward-looking, utopian vision would place the climate action movement in a plane where it can be viewed differently, hence assist in catalyzing and building new identities. It allows the Movement to offer concrete global-scale analysis on how to confront the realities of global power politics, to displace the hegemonic powers of capitalism, and to make justice more central, rather than peripheral, in campaigns. The histories under study have shown how utopian thinking could power people's desire for freedom, equality, and democracy—novelties and dreams that to its many actors were far beyond their initial motivations. Utopian thinking spurs disappointments in the present situation, prods the public to be angry with the current regime, and triggers disruptive transformation and militant emancipation.

In the case of a utopian vision of a 100% renewable energy future, the Movement needs to focus not only the climate ends of the transition but also the complementary visions of decent work (including the vision of a world without work), reduced inequality, and the sense of ownership, justice, and happiness. Since disappointment triggers dissent, it is also important that the Movement underlines and highlights what the barriers are to these visions of positive, prosperous, more durable, and desirable futures. Indeed, the histories under study offered, during their respective times, not only visions of a better future but also effectively tied them to radical critiques of the existing structures of oppression, while remembering their past and present struggles. In the case of the climate action movement, the opponents are easily identifiable: those who fuel climate denialism in the fossil fuel regime complex. These include not only fossil fuel companies but also myopic public policymakers and unsupportive mass media.

Visioning and identity building in the climate action movement has to travel beyond initiating pockets of climate actions toward creating an imaginary that would transform the hegemony that is ordering contemporary human society. It is key to note that the fossil fuel regime complex is but a material manifestation of a social and cultural hegemony that is neoliberal capitalism. It is the neoliberal capitalist hegemon that provides the fossil fuel complex its power (Malm 2016; Battistoni 2018). Addressing this hegemon—following Gramsci's conception of hegemony—would require the Movement to embark on:

- A profound and organic intellectual transition through an educational campaign to instruct new generations what values of sustainability and justice *ought* to dominate future societies;
- The popularization of new cultural codes that imbues sustainability values with equity, fairness, and justice on its core;

- A shift of economic and institutional arrangements that cracks the fossil fuel regime complex and opens up onto a new, malleable, and reflexive, yet focused, sustainability-based regimes; and
- The deployment of infrastructural architectures that repurposes fossil fuel-based infrastructures while pursuing biomimicry, sustainable, and passively designed infrastructures.

The Movement needs to embrace this hegemonic campaign and translate this hegemony into practical and doable solutions through nonviolent tactics. Importantly, the Movement should emphasize that its campaigns are not simply dispersed, autonomous, excessively localized and particular but that these are also about complex and messy politics. It is indeed key to note that this hegemonic campaign is not about a system of domination but as a complex mode of power, emerging from interactions, interconnections, practices, experiences, experimentations, and reflections of heterogeneous and plural groups, organizations, and institutions within society.

A new climate action campaign will start by educating people of the alternatives to neoliberal capitalism, especially the premium it has been giving toward formal modeling approaches over qualitative and sociological epistemologies. This campaign requires more pluralism in public education, bringing, for instance, alternative and new economics to mainstream understanding (e.g. New Economics Foundation, Post-Crash Economic Society, Rethinking Economics, New Systems Project). The Movement should develop meaningful and desirable economic programs that can elaborate alternatives to the neoliberal, fossil fuel-powered capitalism. These alternatives have to be carefully thought out through societal transformations in the age of big data and information, automation, autonomous mobility, cryptocurrencies, post-work, and so on—and not only through the lenses of sustainable energy transition. Economic literacy—i.e. making the "new" economy intelligible to the public at large—is essential in climate actions in ways that economic analyses need to be connected to the insights of everyday, mundane living (cf. Delina and Sovacool 2018).

Instigating cracks in the dominant, singular, and intense view of economics—that is based and promoted on free-market principles—have to be foremost among the strategies for climate actions. Carbon-based infrastructures (e.g. individual and private car-based living), hence, need redesign and repurposing. Since these infrastructures have been shaping individual-focused, carbon-oriented societies regardless of what individuals or communities really want, these structures need to be repurposed immediately.

Future innovation has to be nudged toward the normative ideals of sustainable, less-carbon-based economies that will emphasize which technologies to develop and deploy and who should own and manage them. This direction recognizes that the Movement also has to strategize how it can build a constituency around new identities beyond the Movement's own and present constituencies. This means that the beliefs, desires, behaviors, and perspectives of "the other"—actors in the fossil fuel regime complex such as climate deniers, unsupportive media, myopic governments, fascists, racists, sexists, and capitalists—needs to be nudged toward change.

Since social and political intentions are closely welded into the processes of technological choice (Jasanoff 2004), the Movement needs to emphasize a democratic control over the design, deployment, operations, and management of climate action technologies (Burke and Stephens 2017) and a just transition toward these systems (Delina and Sovacool 2018). Considering that most innovations are publicly funded—from the Internet to green technology to nanotechnology—and that capitalist markets tend toward short-termism and low-risk solutions, public funding for climate action technologies should be made under democratic control (Alperovitz et al. 2017; Hanna 2018; Delina 2016). In that regard, the Movement should work toward having forward-focused, vision-oriented governments that will support global-scale climate through the rapid expansion of cheap, reliable, and sustainable energy systems. State-led mobilizations for the Second World War provides one model by which governments could lead in deploying resources, finance, and institutions to quickly usher in low-carbon-based societies (Delina 2016). The Green New Deal is another variant (New Economics Foundation 2008).

The potential of democratically controlled climate actions—superintended by state planning—goes beyond the rate by which climate action technologies are deployed. The speed of deployment will also include the direction of deployment since effective climate actions would require projects that do not just require marginal improvements to existing technologies but entail entirely new paths of technologies. Think of renewable energy generation, storage, and distribution, for example, which are predominant climate action technologies. In future energy systems, this chain would include an assembly of not only technologies and systems far different from our traditional imageries of grids and utilities but also large-scale public funding that would enable greater public participation and democratic control since almost everyone will be part owners of those new systems (Burke and Stephens 2017; Szulecki 2017).

One way to envisage this is to highlight a role for the Movement pushing an agenda that would strengthen the regulatory powers of the state over industries responsible for climate change with an end-view of enabling stronger public participation in the transition. This would mean opening up electricity markets to community ownership where households, community organizations, small businesses, and local governments become shareholders of the entire energy generation and distribution chain (Gui and MacGill 2018). Radically, it could also mean regaining state control over industries such as electricity utilities—and not merely influencing their decisions with carrots-and-sticks (Alperovitz et al. 2017; Hanna 2018). A re-nationalization program of the electricity industry, including opening it up to control by its workers, would also be ideal (e.g. Aunphattannasilp 2018; Price 2018; Quiggin 2017) but has to be done with extreme caution (e.g. Isaacs and Molnar 2016) and a better appreciation of contexts (Haselip and Potter 2010). Internationally, the Movement could also push for a fossil fuel non-proliferation treaty to signal a rapid transition (Simms and Newell 2018).

The climate action movement, however, needs to distinguish between climate action technologies that it chooses to support. The Movement has to establish clear criteria to adjudicate the potentials of these proposals against their context-specific use. If these technologies—including renewables—will only mean exploiting workers, then its role in the future has to be critically problematized. Nuclear energy—which remains financially risky (Gilbert et al. 2017), has inherent safety risks (Wheatley et al. 2016) and can be exploited to inflict mass destruction (Khan 2017)—and geoengineering—which carries essential governance risks (Szerszynski et al. 2013)—should also be critically challenged.

One thing is certain, and the Movement needs to pay close attention to this: without shifts in hegemonic ideas of society, climate action technologies can remain beholden to neoliberal capitalist values. This is where the lessons of Indian and Philippine examples of mobilizations—where regimes would buy in some time negotiating for some relief to delay, if not prevent, real transformations—are instructive. The Movement has to stick to its work, that is, not to demonstrate to the fossil fuel regime complex that they were wrong but to force them out.

In other words, the climate change is not a market failure to be fixed or reformed but a market system failure to be replaced and a system to be transformed.

REFERENCES

Ackermann, P., & Du Vall, J. (2000). *A Force More Powerful: A Century of Nonviolent Conflict.* New York: St. Martin's Press.

Aiken, G. (2012). Community transitions to low carbon futures in the transition towns network (TTN). *Geography Compass, 6,* 89–99.

Alperovitz, G., Guinan, J., & Hanna, T. M. (2017, April 26). The policy weapon climate activists need. *The Nation.*

Aquino, B. A. (1984). Political violence in the Philippines: Aftermath of the Aquino assassination. *Southeast Asian Affairs,* 266–276.

Aunphattannasilp, C. (2018). From decentralization to re-nationalization: Energy policy networks and agenda setting in Thailand (1987–2017). *Energy Policy, 120,* 593–599.

Barr, S., & Devine-Wright, P. (2012). Resilient communities: Sustainabilities in transition. *Local Environment, 17,* 525–532.

Battistoni, A. (2018, August 3). How not to talk about climate change. *Jacobin.* https://bit.ly/2Dwn4Br.

Bennett, W. L., & Segerberg, A. (2012). The logic of connective action: Digital media and the personalization of contentious politics. *Information, Communication & Society, 15,* 739–768.

Bomberg, E., & McEwen, N. (2012). Mobilizing community energy. *Energy Policy, 51,* 435–444.

Boudreau, V. (2004). *Resisting Dictatorship: Repression and Protest in Southeast Asia.* New York: Cambridge University Press.

Bradshaw, E. A. (2015). Blockadia rising: Rowdy greens, direct action and the Keystone XL pipeline. *Critical Criminology, 23,* 433–448.

Bulkeley, H., Castan-Broto, V., Hodson, M., & Marvin, S. (2010). *Cities and Low Carbon Transitions.* London: Routledge.

Burke, M. J., & Stephens, J. C. (2017). Energy democracy: Goals and policy instruments for sociotechnical transitions. *Energy Research & Social Science, 33,* 35–48.

Chakrabarty, B. (1997). *Local Politics and Indian Nationalism, Midnapur 1919–1944.* New Delhi: Mahonar Publishers & Distributors.

Chatterton, P. (2013). Towards an agenda for post-carbon cities: Lessons from Lilac, the UK's first ecological, affordable cohousing community. *International Journal of Urban and Regional Research, 37,* 1654–1674.

Connors, P., & McDonald, P. (2011). Transitioning communities: Community, participation and the transition town movement. *Community Development Journal, 46,* 558–572.

Dalton, D. (1993). *Mahatma Gandhi: Nonviolent Power in Action.* New York: Columbia University Press.

Delina, L. L. (2016). *Strategies for Rapid Climate Mitigation: Wartime Mobilisation as a Model for Action?* Abingdon: Routledge.

Delina, L. L. (2018). Energy democracy in a continuum: Remaking of public engagement on energy transitions in Thailand. *Energy Research & Social Science, 42,* 53–60.

Delina, L. L., & Sovacool, B. K. (2018). Of temporality and plurality: An epistemic and governance agenda for accelerating just transitions for energy access and sustainable development. *Current Opinion in Environmental Sustainability, 34,* 1–6.

Della Porta, D., & Diani, M. (2006). *Social Movements: An Introduction* (2nd ed.). Malden: Blackwell.

Diani, M. (1992). The concept of social movement. *The Sociological Review, 40,* 1–25.

Dietz, M., & Garrelts, H. (Eds.). (2014). *Routledge Handbook of the Climate Change Movement.* Abingdon: Routledge.

Escobar, A., & Alvarez, S. E. (Eds.). (1992). *The Making of Social Movements in Latin America: Identity, Strategy, and Democracy.* New York: Routledge.

Fossil-Free. (2018). *Divestment Commitments.* https://gofossilfree.org/divestment/commitments/.

Foxon, T. (2013). Transition pathways for a UK low carbon electricity future. *Energy Policy, 52,* 10–24.

Fredrickson, G. M. (1995). *Black Liberation: A Comparative History of Black Ideologies in the United States and South Africa.* Oxford: Oxford University Press.

Gamson, W. A. (1991). Commitment and agency in social movements. *Sociological Forum, 6,* 27–50.

Gilbert, A., Sovacool, B. K., Johnstone, P., & Stirling, A. (2017). Cost overruns and financial risk in the construction of nuclear power reactors: A critical appraisal. *Energy Policy, 102,* 644–649.

Gui, E. M., & MacGill, I. (2018). Typology of future clean energy communities: An exploratory structure, opportunities, and challenges. *Energy Research & Social Science, 35,* 94–107.

Hanna, T. M. (2018). *Our Common Wealth: The Return of Public Ownership in the United States.* Manchester: Manchester University Press.

Hardiman, D. (1981). *Peasant Nationalists of Gujarat: Kheda District, 1917–1934.* Oxford: Oxford University Press.

Haselip, J., & Potter, C. (2010). Post-neoliberal electricity market 're-reforms' in Argentina: Diverging from market prescriptions. *Energy Policy, 38,* 1168–1176.

Hetherington, K. (1998). *Expressions of Identity: Space, Performance, Politics.* London: Sage.

Hodges, H. E., & Stocking, G. (2015). A pipeline of tweets: Environmental movements' use of Twitter in response to the Keystone XL pipeline. *Environmental Politics, 25,* 223–247.

Hopkins, R. (2008). *The Transition Handbook: From Oil Dependency to Local Resilience.* Totnes: Green Books.

Isaacs, R., & Molnar, A. (2016). Island in the neoliberal stream: Energy security and soft re-nationalisation in Hungary. *Journal of Contemporary European Studies, 25,* 107–126.

Jacobson, M., Delucchi, M., Bauer, Z., Goodman, S., Chapman, W., Cameron, M., et al. (2017). 100% clean and renewable wind, water, and sunlight all-sector energy roadmaps for 139 countries of the world. *Joule, 1,* 108–121.

Jasanoff, S. (Ed.). (2004). *States of Knowledge: The Co-production of Science and Social Order.* London: Routledge.

Khan, S. (2017). *Nuclear Proliferation Dynamics in Protracted Conflict Regions: A Comparative Study of South Asia and the Middle East.* London: Routledge.

King, M. L., Jr. (2010 [1958]). *Stride Toward Freedom: The Montgomery Story.* Boston: Beacon Press.

Levitas, R. (2011). *The Concept of Utopia.* Oxford: Peter Lang.

Lichterman, P. (1999). Talking identity in the public sphere: Broad visions and small spaces in sexual identity politics. *Theory and Society, 28,* 101–141.

Llewelyn, D. H., Rohse, M., Day, R., & Fyfe, H. (2017). Evolving energy landscapes in the South Wales Valleys: Exploring community perception and participation. *Energy Policy, 108,* 818–828.

Lorenzoni, I., Nicholson-Cole, S., & Whitmarsh, L. (2007). Barriers perceived to engaging with climate change among the UK public and their policy implications. *Global Environmental Change, 17,* 445–459.

Malm, A. (2016). *Fossil Capital: The Rise of Steam Power and the Roots of Global Warming.* London: Verso.

Masani, Z. (1987). *Indian Tales of the Raj.* Berkeley: University of California Press.

Masselos, J. (1993). *Indian Nationalism: An History.* New York: Sterling Publishers.

McKenna, R., Merkel, E., & Fichtner, W. (2016). Energy autonomy in residential buildings: A techno-economic model-based analysis of the scale effects. *Applied Energy, 189,* 800–815.

McKibben, B. (2013, February 22). The case for fossil-fuel divestment. *Rollingstone.*

Melucci, A. (1989). *Nomad of the Present: Social Movements and Individual Needs in Contemporary Society*. Philadelphia: Temple University Press.

Melucci, A. (1996). *Challenging Codes: Collective Action in the Information Age*. Cambridge: Cambridge University Press.

Mercado, M. A. (Ed.). (1986). *People Power: An Eyewitness History, the Philippine Revolution of 1986*. Manila: Writers and Readers Publishing with Tenth Avenue Editions.

Mey, F., & Diesendorf, M. (2018). Who owns and energy transition? Strategic action fields and community wind energy in Denmark. *Energy Research & Social Science, 35*, 108–117.

Moane, G. (2011). *Gender and Colonialism: A Psychological Analysis of Oppression and Liberation*. Houndmills: Palgrave Macmillan.

Moser, S. C. (2009). Costly knowledge—Unaffordable denial: The politics of public understanding and engagement on climate change. In M. T. Boykoff (Ed.), *The Politics of Climate Change: A Survey* (pp. 155–181). Oxford: Routledge.

Müller, M. O., Stämpfli, A., Dold, U., & Hammer, T. (2011). Energy autarky: A conceptual framework for sustainable regional development. *Energy Policy, 39*, 5800–5810.

Nehru, J. (2005 [1958]). *A Bunch of Old Letters*. New Delhi: Viking.

Nepstad, S. E. (2011). *Nonviolent Revolutions: Civil Resistance in the Late 20th Century*. Oxford: Oxford University Press.

New Economics Foundation (NEF). (2008). *A Green New Deal: Joined-up Policies to Solve the Triple Crunch of the Credit Crisis, Climate Change and High Oil Prices*. London: NEF.

Overholt, W. H. (1986). The rise and fall of Ferdinand Marcos. *Asian Survey, 26*, 1137–1163.

Paliewicz, N. S. (2018). Making sense of the people's climate march: Towards an aesthetic approach to the rhetoric of social protest. *Western Journal of Communication*. https://doi.org/10.1080/10570314.2018.1446548.

Pearce, W., Brown, B., Nerlich, B., & Koteyko, N. (2015). Communicating climate change: Conduits, content, and consensus. *WIREs Climate Change, 6*, 613–626.

Price, S. (2018). Majority agree: Renationalise electricity. *Green Left Weekly, 1188*, 3.

Quiggin, J. (2017, March 5). The case for renationalising Australia's electricity grid. *The Conversation*.

Quitzow, L., Canzler, W., Grundmann, P., Leibenath, M., Moss, T., & Rave, T. (2016). The German Energiewende—What's happening? *Utilities Policy, 41*, 163–171.

Saunders, C. (2008). Double-edged swords? Collective identity and solidarity in the environment movement. *The British Journal of Sociology, 59*, 227–253.

Schock, K. (2005). *Unarmed Insurrections: People Power Movement in Nondemocracies*. Minneapolis: University of Minnesota Press.

Segerberg, A., & Bennett, W. L. (2011). Social media and the organization of collective action: Using Twitter to explore the ecologies of two climate change protests. *The Communication Review, 14*, 197–215.

Seyfang, G., Park, J. J., & Smith, A. (2013). A thousand flowers blooming? An examination of community energy in the UK. *Energy Policy, 61*, 977–989.

Sicotte, D. M., & Joyce, K. A. (2017). Not a 'petro metro': Challenging fossil fuel expansion. *Environmental Sociology, 3*, 337–347.

Simms, A., & Newell, P. (2018, October 23). We need a fossil fuel non-proliferation treaty—And we need it now. *The Guardian*.

Snow, D. A. (2013). Framing and social movements. In D. A. Snow, D. della Porta, B. Klandermans, & D. McAdam (Eds.), *The Wiley-Blackwell Encyclopedia of Social and Political Movements*. Malden, MA: Blackwell.

Srnicek, N., & Williams, A. (2016). *Inventing the Future: Postcapitalism and a World Without Work*. London: Verso.

Stryker, S., & Burke, P. J. (2000). The past, present, and future of identity theory. *Social Psychology Quarterly, 63*, 284–297.

Szerszynski, B., Kearnes, M., Macnaghten, P., Owen, R., & Stilgoe, J. (2013). Why solar radiation management geoengineering and democracy won't mix. *Environment and Planning A: Economy and Space, 45*, 2809–2816.

Szulecki, K. (2017). Conceptualizing energy democracy. *Journal of Environmental Politics, 27*, 21–41.

Van der Schoor, T., & Scholtens, B. (2015). Power to the people: Local community initiatives and the transition to sustainable energy. *Renewable and Sustainable Energy Reviews, 43*, 666–675.

Wang, S., Corner, A., Chapman, D., & Markowitz, E. (2018). Public engagement with climate change imagery in a changing digital landscape. *WIREs Climate Change, 9*, e509.

Weber, T. (1997). *On the Salt March*. New Delhi: HarperCollins.

Wheatley, S., Sovacool, B. K., & Sornette, D. (2016). Of disasters and dragon kings: A statistical analysis of nuclear power incidents and accidents. *Risk Analysis, 37*, 99–115.

Yagatich, W., Galli Robertson, A. M., & Fisher, D. R. (2018). How local environmental stewardship diversifies democracy. *Local Environment: The International Journal of Justice and Sustainability, 23*, 431–447.

Zunes, S. (1999). The origins of people power in the Philippines. In S. Zunes, S. B. Asher, & L. L. Kurtz (Eds.), *Non-Violent Social Movements: A Geographical Perspective* (pp. 129–157). Malden: Blackwell.

CHAPTER 4

Culturing and Framing: Working on the Ills of the Past, in the Present, for Tomorrow's Benefits

Abstract The proliferation of climate-related images and symbolisms, such as pictures of emasculated bears, drowned cities, and polluted air, had led to some climate action but was weakly received in general. This weak reception occurs for myriad reasons. The behavioral and brain sciences suggest that the human moral judgment system is poorly equipped to identify future, large-scale, and long-term hazards such as those brought about by climate impacts. Culturing and framing are essential in effective climate action communication and require a better understanding of human behaviors. Visceral experiences of trigger events, mostly in the form of some psychological tipping points remain important; but they are never guaranteed. Campaigns where individuals could care about a sustainable and just future and empathize with people affected by climate impacts are the ones that are well received. Culturing and framing, thus, would involve narratives of hope, the sense of the possible, pride, and gratitude that could be attached to many climate actions in spaces where people can contribute using their existing capacities.

Keywords Culturing · Framing · Climate denial · Climate symbols · Dissonance · Climate porn · Climate justice · Democracy

One of the biggest challenges faced by all social movements is how to make sense of doing "what is right" felt rapidly across all sectors of society. The contemporary climate action movement is no exception.

© The Author(s) 2019   53
L. L. Delina, *Emancipatory Climate Actions*,
https://doi.org/10.1007/978-3-030-17372-2_4

With behavioral and brain sciences suggesting that the human moral judgment system is poorly equipped to identify future, large-scale, and long-term hazards such as climate impacts (Swim et al. 2011; Weber 2006), messaging climate actions—working on the ills of the past in the present for tomorrow's benefits—has become more difficult. Our evolution has impeded our capacity to react today on the future ramifications of our historical and current actions (Gifford 2011; Gifford et al. 2018). Instead of doing climate actions now, denial has been the standard response (Oreskes et al. 2008).

Denying climate actions does not only involve deny ing the science of climate change per se. In its conventional sense, denial can be active, that is, being knowledgeable of the facts but refusing to acknowledge them, while refuting any possible response to the problem because these responses threaten their income, profession, status, and values (Stoknes 2014; Dunlap and McCright 2011). Denial, can also be passive, that is, being indifferent or unresponsive. One may know but does not care, opting silence over action, restricting exposure to inconvenient facts, and ignoring the necessary solutions (Stoknes 2014).

Denial, thus, takes a variety of forms. It ranges from ignoring the facts ("I don't know the consequences of my actions.") to rejecting blame ("I have done nothing wrong.") to condemning the accuser ("You have no right to challenge me.") to denying responsibility ("I do not cause this problem.") to powerlessness ("I am but a speck in the universe.") to comfort ("It is too difficult to change my behavior.") to fabricating constraints ("There are too many challenges and they are huge.") (Stoll-Kleeman et al. 2001; cf. Björnberg et al. 2017). In short, deniers advance what can be called "cockroach" ideas—false claims that you may think you have gotten rid of, but keep coming back (Krugman 2018).

The climate action movement addresses denial usually by employing stimulating messages found in images, symbols, languages, and emotions. Pictures of animals, such as an emaciated polar bear standing forlorn on thinning ice, typically supplement climate change information with some appeal to human emotions. For some time, this messaging approach would attract an evocative response. Over time, however, the overuse of polar bears and other symbols numbed the public (Gifford 2011; Lewandowsky and Whitmarsh 2018). The same psychological shifts happened with images of sinking cities and human deaths from extreme weather events. While these messages may fuel emotions, they could also induce unease, fear, and/or guilt (Bain et al. 2012; Chapman et al. 2017).

These types of messages failed to reach their full potential because they were not matched with actual climate actions. In short, there was cognitive dissonance (Stoknes 2014; Kraft et al. 2015).

Dissonance suggests that one can always change the interpretation of their own action if they failed to change action (Festinger 1962). "Those grapes surely must be sour," says the fox after coveting inaccessible grapes. Similarly, one could say climate actions are not that important since we are not doing it effectively anyway. The ways by which many modern, rich and Western human societies are ordered and built—around individualism, independent driving, suburbia, energy-intensive agriculture, meat eating, air travel, globalization—are in complete tension with what we already know to be the most rational climate action. Instead of addressing the root causes of climate change, our responses, so far, have centered around coping and adapting. Our current dissonant response needs to change—and this process requires coherent and clear messaging tools.

Humans do not automatically or naturally attach meanings to objects, events or experiences that they encounter; instead, meanings arise through interpretive processes that are mediated by **culture** (Snow 2013; Kurzman 2008; cf. Crane 1994). How we view ourselves and how others perceive us are affected by culture. For effective climate actions, such as the rapid 100% energy transitions, circular economy, etc., to be effectively communicated, Andy Stirling (2014) suggests to look beyond scientific language, policy prescription, and technological deployment and to consider "culturing."

To Stirling, culturing involves the patterns and practices by which social agency and structure are implicated in production and consumption. It is, Stirling further argues, essential that this interpretive work is complemented with proper, well-designed, and effective **framing**. Framing first focuses our attention; it delineates what is relevant and what is not. Framing then assists us to articulate; it helps us tie elements so that a specific storyline rather than another is told. Framing finally contributes to our transformative exercises; it determines how we mobilize our grievances (Snow 2007; Benford and Snow 2000).

Culturing and framing—essential in effective climate action communication—require, front-and-center, a better understanding of human behaviors. In histories under study, people mobilized, with passion and determination, following their visceral experiences of trigger events, mostly in the form of some psychological tipping points (Cuny 1983; Pelling and Dill 2009). Often, these tipping points were achieved when people started realizing, strongly and powerfully, that their deeply held values were being violated (Markowitz and Shariff 2012; Swim et al. 2011). Movements strategically used these moments to communicate to people the moral violations of their opponents, and effectively used these moments in culturing and framing new messages that would ignite large-scale mobilizations.

CULTURING AND FRAMING IN YESTERDAY'S SOCIAL ACTIONS

The arrest of Rosa Parks became a symbol that effectively represented the racial injustice of her time. Arresting someone with "unblemished reputation" and one who is a "good citizen" (Boyle-Baise 2003) momentously tipped the social balance in her favor. Parks did conform to the type of an African-American who was deserving of a right to a bus seat she herself paid for. Her character was "pristine;" she was "humble and gentle" (Harrington 2000; Parks 1992). Her behavior as a "well-mannered person," her "serene demeanor," her "proper speech," her "humble, saintly way," her "ascetic lifestyle" carried not only the image but also the reality of an African-American deserving of respect and the right to live peacefully in her community (Harrington 2000). Edgar Daniel Nixon, who was the president of the Montgomery chapter of the National Association for the Advancement of Colored People, said in a media interview following the arrest: "She was honest, she was clean, she had integrity. The press couldn't go out and dig up something she did last year, or last month, or five years ago. They couldn't hang nothing like that on Rosa Parks" (Parks 1992: 125).

Rosa Parks epitomized an African-American's, or any person's for that matter, right to live in Montgomery. No rational person, even a white person, could argue that she did not deserve a seat on her bus ride home from work that December evening. Parks, and her life, effectively and successfully provided an important, visible and, most importantly, proximate moral symbol and frame for her Movement. She became an icon that the civil rights group in Montgomery direly needed at the time and had been patiently waiting for (Parks 1992). Her arrest shocked her community and sent an effective message that eventually propelled the modern movement for civil rights in America.

In the Indian case, an upright guru and his simple act of salt-making provided the necessary symbol that effectively framed the Movement for Free India. Gandhi successfully conjured symbols that are attached to India's religions: that of the great warrior Lord Rama "on his way to conquer Sri Lanka," of the Buddha "inspired by the mission of relieving the grief-stricken and downtrodden" (Dalton 1993: 109), and of the Prophet Mohammad who leads from darkness to light. Gandhi translated these symbols into frames and messages of nonviolent dissent in his campaigns.

In the Philippines, the assassination of Ninoy Aquino provided the frame necessary to spark Filipinos to rise to the occasion and to dissent Marcos's corrupt regime. The framing of Ninoy's assassination powerfully resonated on the country's devoutly Catholic population (Schock 2005). Jaime Cardinal Sin, the archbishop of Manila, condemned the murder in the strongest terms and declared the slain leader "a national martyr" (cf. Thompson 1995). Such virtuous and saintly characterization of a person was once again used to define the choice of Cory Aquino, Ninoy's widow, as the presidential candidate of the united opposition during the snap elections of 1986. Compared to the image of Marcos as a traditional and corrupt politician, Cory provided a new image of a totally inexperienced politician (she was a housewife) but a person untarnished by any charge of corruption. She was the personification of a luminous and cheerful virtue, comparable to the image projected by Rosa Parks in Montgomery and Gandhi during his salt march, against the dark and the brooding image of a corrupt Ferdinand Marcos. During the campaign for the snap election, Cory would frequently invoke this framing, often telling the crowd her late husband's words: "We cannot fight Marcos with arms, because he has so many. We could not fight him with money, because we do not have any. The only way we can fight him is with morality" (Komisar 1987: 78).

The symbolism, charged with moral virtues and personified by Cory, echoed throughout the Philippines and eventually found a home in many Filipino hearts. This process of culturing led ordinary Filipinos to contribute what money they could to support the "good" widow's campaign. Filipinos would put up homemade posters to signify their support for her and their dislike of the dictator (David 1996). The 1986 election, according to Filipino sociologist Randolf David (1996: 8), was "nothing less than a battle between good and evil. It was no longer just a political contest. This was a moral war."

The symbolisms brought about by the "good" personas and "upright" moralities of Mahatma Gandhi, Rosa Parks, and Cory Aquino (and also by her martyred husband, Ninoy), along with the events that highlighted their moral ascendancies, harmoniously connected with the aspirations of the subjugated Indian masses, African-Americans, and the

Filipino people of their respective times. These moral symbolisms were important, but, on their own, are not strong enough to extend campaigns such as one-day boycotts, demonstrations, rallies, or strikes into year-long (many times, even longer) campaigns and propel social action movements to greater heights. What the histories under study suggest is to employ these symbols not as an end in themselves but as means to frame new messages; in short, to raise a new public culture. It is only until then that Movements had become self-perpetuating forces.

CULTURING AND FRAMING IN TODAY'S CLIMATE ACTIONS

Many in the climate action movement understand the connection between symbols, frames, and messages (cf. Moser 2010). Psychological stimuli have been duly tapped to hinge many past and present climate action strategies such as by using vocabularies related to shocking events, including even those that are yet to occur (e.g. Wallace-Wells 2017). The assumption for this nihilistic framing rests on the idea that shocking events, especially those that can be personally witnessed, if not experienced, could be used as climate action triggers (Demski et al. 2017; cf. Leiserowitz and Smith 2017; Spence et al. 2011) or, at the very least, attract attention to climate change (Sisco et al. 2017). These messages also involve frames that bank on exogenous experiences, using, for example, extreme weather events that occurred in distant lands and their impacts to people's lives and livelihoods to spur an audience onto action. Overall, these framings would contain images, languages, and symbols designed to appeal on emotions (cf. Leiserowitz and Smith 2017). Examples include images of an emasculated polar bear stranded on a thin, breaking ice; of submerged coastal cities; of a graph of hockey stick-like, fast-rising emission trends; and vivid descriptions of "uninhabitable Earth:" famine, economic collapse, and a sun that cooks us (Wallace-Wells 2017). This choice of doomsday images and symbols is an attempt to simulate the viewing, reading, or listening audience to reach an emotional "high" in the hope that they will then engage with climate actions (Stern 2012; cf. Roberts 2017). The hope is that these narrations and pictures will result in anger, which could then motivate people to rectify injustices (Thomas et al. 2009; cf. Bain et al. 2012).

An audience's perception of these frames and messages, however, varies at best (Chapman et al. 2017). Emotionally-charged burden framing may grab some people's attention and may even mobilize some to adopt climate actions. A focus on spatially and temporally distant events, technical languages, and fear appeals, however, can also backfire and disengage some audiences (Mann et al. 2017; Weber 2010; Moser 2007). An audience may also perceive these messages as a form of manipulation (Lorenzoni et al. 2007; O'Neill and Nicholson-Cole 2009). Despite the noble intention of these messaging tools, they rarely push people to action (Stevenson and Peterson 2015).

Weak reception occurs for myriad reasons, but the behavioral and brain sciences explain that the human moral judgment system is but poorly equipped in identifying future, large-scale, and long-term hazards such as those brought about by climate impacts (Swim et al. 2011). The evolutionary history of the human species simply impedes our capacity to react today on the future ramifications of our historical and present actions (Gifford 2011). Psychological repercussions, such as denial and apathy, also render burden framing of limited value in mobilization (American Psychological Association 2009; Center for Research on Environmental Decisions 2009). Another issue with too much reliance on burden frames is that climate actions could not be guaranteed even the Movement is granted similar pivotal events (e.g. Typhoon Haiyan, California wildfires, etc.). Such extreme events—although some already manifest—also occur too gradually to be effectively used as narratives to drive effective climate actions. Complementing climate action messages with symbols and frames that would offer hope, bear solutions, and that are broadly consistent with the personal interests, aspirations, hopes, desired identities, and cultural biases of specific audiences tend to be more appealing; hence, such messages are the best received (Bain et al. 2012; Weber 2010; Segnit and Ereaut 2007; O'Neill and Nicholson-Cole 2009). Connecting climate actions to people's own experiences and encounters is key.

Culturing and Framing in Tomorrow's Climate Actions

The most engaging images of social action are not those that convey fear and extraordinary burden, but the optimistic icons that narrate effective solutions, that would make people feel that they are able to do something about the issue, and, at the same time, see personal benefits from doing or contributing to that action (Bain et al. 2012; O'Neill and Nicholson-Cole 2009; American Psychological Association 2009; Center for Research on Environmental Decisions 2009; Kahan and Braman 2008).

Proximity to symbols of hope, in terms of closeness to an audience's own value systems, mundane lives, experiences, and local area, is a lesson that the histories under study have illustrated. Indians found it in Gandhi (and in his nonviolent discipline). People in Montgomery found it in Rosa Parks (and in her virtue filled life). Filipinos found it in Ninoy Aquino (and in his self-sacrifice). People found hope through clearly demonstrated, available, and feasible alternatives to the regimes of their time.

The gap between effective climate actions and their truly measurable effects (which only accrue further into the future) generally betrays the climate action movement with an argument for effective climate actions in the present. Since climate actions today will largely be for the benefit of future generations, the Movement needs to develop strategies that would focus some forms of benefits of climate actions into the present, while acknowledging that this generation is already locked into a world of weather extremes. Motivating people for action can be best achieved using strategies that would directly affect an audience's personal, lived experiences (Fisher and McInerney 2012; Bain et al. 2012). Climate actions, thus, have to move toward translating "the mundane politics of everyday life, into a directly embodied political process of movement mobilisation for a genuine strategy for transformation" (Rosewarne et al. 2014: 17), or, in gist, towards the "culturing" of transformations (Stirling 2014).

Invoking local and current rationales for climate actions must also be broadly consistent with the aspirations, desired social identities, and cultural biases of individuals where climate action messages are directed at (Fisher and McInerney 2012; Weber 2010; Segnit and Ereaut 2007; Markowitz and Shariff 2012; Swim et al. 2011; Haidt 2007). It is key that climate actions be oriented and connected to the values that a specific audience holds on to dearly (O'Neill and Hulme 2009; Weber 2006; American Psychological Association 2009; Lorenzoni et al. 2007). Indeed and in some instances, it would seem paradoxical that climate action messages must be conveyed to some groups and communities not necessarily with scientific facts (which these groups would not even listen or pay attention to) to convince but with their core values so as to give climate facts and climate actions a fighting chance (Mooney 2011; cf. Kahan et al. 2011). This is particularly true in the United States and Australia where the climate action message becomes submerged by left–right politics: an imbalance between those prioritizing actions who support higher taxes and regulations on energy and carbon emissions and those whose political values prioritize lower taxes and less regulation. With climate actions in these two high-emission countries effectively attached by conservatives with the political identity of the "liberals," such that it has now become the cultural identity of a US Republican conservative to oppose climate actions, if not to deny climate change at all, it will be essential for the Movement to creatively design their messages using prompts and frames that motivate these sections of society into action.

Greater emphases and efforts to underline positive and values-based solutions should move the climate action movement from its overemphasis on "diagnostic frames" toward "motivational frames" (cf. Snow and Benford 1988). Rather than invoking distant and complex messaging tools, which have been proven to induce guilt, shame, and/or anxiety (Markowitz and Shariff 2012), the Movement should deploy campaigns where individuals could care about a sustainable and just future and empathize with people affected by climate impacts. In framing present benefits of climate actions, the Movement can adopt—following thorough context-specific analyses—narratives of hope, the sense of the possible, pride and gratitude attached to many climate actions. This would mean that campaigners have to carefully weave realistic and credible stories about the repercussions of unabated emissions with positive discourses on the contributions of climate actions on people's jobs, income, savings, health, sustainability, democracy, empowerment, etc.

Bringing climate actions to spaces where people could participate and engage directly is key. Local climate actions—in households, neighborhood associations, church communities, offices, factories, schools, towns, municipalities, colleges, universities, cities, etc.—have already been providing optimistic climate solutions. These climate actions include: adoption of rooftop household and community solar, wind, or hydro power; energy efficient school, college, university, and community buildings; friendly pedestrian walkways and bikeways; improved public transport systems; community energy cooperatives; green purchasing; etc. These are visible messaging tools that highlight and demonstrate the alternatives to a fossil fuel-based society first-hand. They also offer direct and indirect benefits to people in terms of monetary savings, financial rewards, opportunity to know and meet neighbors and colleagues, etc. These climate actions also open up opportunities for people who want to participate but do not necessarily see themselves involved in street demonstrations, for example.

Nonviolent protests against the fossil fuel regime complex, nonetheless, remain essential tools and paramount tactics for climate action.

These include dissent against the expansion of engineering and technical systems that perpetuate the regime, such as blockages on oil pipelines and new coal-fired power plant installations (Bradshaw 2015; McKibben 2013), fossil fuel divestment (Fossil-Free 2018), and court cases against regime-holders. Three examples of the latter are seen in the Philippines (where a legal case considers if the emissions from fossil fuel companies violate the human rights of those hit by extreme weather events, see Howard 2016), United States (where a federal lawsuit, *Juliana v. United States*, was filed by 21 youth plaintiffs against the Obama and Trump administrations, see Irfan 2018), and the Netherlands (where the Dutch government was ordered to cut its emissions, see Neslen 2015).

Nonviolent dissent elevates the status of climate change issue in public perception since such actions are the most talked about in mass and social media. Historical mobilizations also suggest that evoking images of nonviolent campaigns could assist in making some disaffected elites quickly switch sides and spurring many ordinary citizens to participate (Nepstad 2011). Nonviolence is only one of the many virtues that climate action campaigners must exemplify; the ways they are framed as articulators of change also matter—as shown, for example, by the untainted personalities of Gandhi, Park, and the Aquinos. The consistency between claims and actions, and the credibility, competence, morality, and warmth of climate action messengers indeed affect the message itself (cf. Snow 2013; Bain et al. 2012; Fiske et al. 2002; Judd et al. 2005; Leach et al. 2007).

Small-scale, community-based, and local actions or folk politics are still essential; however, they are not enough (Srnicek and Williams 2016; cf. Anderson and Bows 2012). The Movement needs to inculcate a new culture by messaging climate actions that push for large-scale social, political, and economic (not just technical) changes. This framing is essential in building a counter movement to the present fossil fuel capital-based hegemony. The time is ripe for the Movement to advance messaging frames that advocate for big, macro-political, and ideological ideas such as post-capitalism, future work, and utopia (see Chapter 3; Srnicek and Williams 2016). Especially in the wake of emerging technologies like autonomous driving, artificial intelligence, and big data, illiberal democracy, and labor and employment risks, a new social contract is now necessary (e.g. The Next Systems Project 2018; Delina 2016; Zilk 2016; Bauhardt 2014; Cattaneo et al. 2012; German Advisory Council on Global Change 2011). This new thrust is salient since people typically want to live in a strongly developed society (in terms of scientific progress and economic growth) with minimal social dysfunctions (such as crime and poverty) (Bain et al. 2012).

The Movement is in position to advance these new frames as illustrated in the histories under study. The Movement for Free India, the American Civil Rights Movement, the Philippine People Power Revolution, and the Burmese Pro-democracy Movement—though spurred and inspired by local-based triggers and their consecutive mobilizations—were (and, inarguably, are) continuing struggles for much bigger ideals of freedom, justice, and democracy. In moving beyond folk politics (which remain essential nonetheless), the climate action movement can adopt the hegemonic idea of sustainability expansion and universal, just emancipation.

Culturing and framing the hegemonic idea of sustainability expansion and universal emancipation has a moral imperative backed up by present violations of peoples' deeply held "good" values. Sometime soon, people will begin realizing that their current state of affairs—inequality, hatred, neocolonialism, racism, misogyny, paranoia, authoritarianism, and unsustainability—transgresses the values that they hold precious (cf. Markowitz and Shariff 2012; Swim et al. 2011; Haidt 2001, 2007). Once these tipping points arrived, change becomes inevitable (cf. Center for Research on Environmental Decisions 2009; Weber 2010; American Psychological Association 2009) and large-scale activism tends to ignite (Moyer et al. 2001).

Within the climate action movement, the moral basis for strong climate actions already exists. Indeed, climate change has been extensively framed as a matter of right and wrong. Bill McKibben, one of the most prominent and influential voices in the Movement, writes: "[The] more carefully you do the math, the more thoroughly you realize that [climate change], at bottom, [is] a moral issue" (McKibben 2012: 5). This moral imperative is also key in setting the normative definition of climate actions—the way the world *ought* to be—and in understanding its temporal terms, both immediate and intergenerational (Rosewarne et al. 2014: 91). The "reconception" of the concepts of justice and democracy holds an important framing device.

Justice can be conceptualized in multiple ways, but central to this concept is the protection of basic needs, rights, and political processes—an elaboration of an alternative to the hegemony presently held by fossil-fueled neoliberal capitalism. A justice-based approach to climate actions advances sustainability expansion with universal emancipation described by fair distributions (what are the concerns of the climate action), political and social recognition (who the climate action affects), and procedural inclusion (how the climate action could be mediated or remediated). Within a justice approach lies procedural inclusion: meaning that climate actions must be clear and explicit and done with active public engagement processes that take note, appreciate, and include the different values, discourses, and imaginaries of potential courses of action. Distributional justice responds to this: what are the distributional burdens of climate actions, and how do we address them. Democratically mobilized climate actions are key in embedding the voices of the traditionally marginalized, ignored, and misrepresented sections of society. A justice-based approach, thus, includes understanding and evaluating: who the ignored actors are and how could they be recognized. Recognition-based justice must move toward ensuring procedural justice. Are the processes fair and what new exercises or processes can be developed to ensure fair processes?

Democratizing climate actions entails creating new systems, processes, and arrangements, including participatory decision-making that lead to inclusive, deliberative, and influential public engagement with climate actions. The Movement should demand climate actions to be conducted in ways that recognize and represent heterogeneous interests, values, and reflections of citizens and non-citizens in respectful and friendly avenues.

Also included in new framing is the creation of new sociotechnical, governance, and economic models that can inform the political goals of the climate action movement, as well as fostering public literacy in the new economics of climate actions (see Chapter 3). Making the "new economy" intelligible to ordinary publics—jargon-free and beyond the confines of research institutes and universities—is democratizing. This requires connecting sophisticated analyses of economic trends with the insights of people's every day, ordinary, and seemingly mundane lives and experiences.

REFERENCES

American Psychological Association (APA). (2009). *Psychology and Global Climate Change: Addressing a Multi-faceted Phenomenon and Set of Challenges. A Report by the Task Force on the Interface Between Psychology and Global Climate Change*. Washington, DC: APA.

Anderson, K., & Bows, A. (2012). A new paradigm for climate change. *Nature Climate Change, 2,* 639–640.

Bain, P. G., Horsey, M. J., Bongiorno, R., & Jeffries, C. (2012). Promoting pro-environmental action in climate deniers. *Nature Climate Change, 2,* 600–603.

Bauhardt, C. (2014). Solutions to the crisis? The Green New Deal, degrowth, and the solidarity economy: Alternatives to the capitalist growth economy from an ecofeminist economics perspective. *Ecological Economics, 101,* 60–68.

Benford, R. D., & Snow, D. A. (2000). Framing processes and social movements: An overview and assessment. *Annual Review of Sociology, 26,* 611–639.

Björnberg, K. E., Karlsson, M., Gilek, M., & Hansson, S. O. (2017). Climate and environmental science denial: A review of the scientific literature published in 1990–2015. *Journal of Cleaner Production, 167,* 229–241.

Boyle-Baise, M. (2003). Doing democracy in social studies methods. *Theory & Research in Social Education, 31,* 51–71.

Bradshaw, E. A. (2015). Blockadia rising: Rowdy greens, direct action and the Keystone XL pipeline. *Critical Criminology, 23,* 433–448.

Cattaneo, C., D'Alisa, G., Kallis, G., & Zografos, C. (Eds.). (2012). Special issue: Politics, democracy and degrowth. *Futures, 44,* 515–654.

Center for Research on Environmental Decisions (CRED). (2009). *The Psychology of Climate Change Communication: A Guide for Scientists, Journalists, Educators, Political Aides, And the Interested Public.* New York, CRED: Columbia University.

Chapman, D. A., Lickel, B., & Markowitz, E. M. (2017). Reassessing emotion in climate change communication. *Nature Climate Change, 7,* 850–852.

Crane, D. (1994). Introduction: The challenge of the sociology of culture to sociology as a discipline. In D. Crane (Ed.), *The Sociology of Culture* (pp. 1–20). Oxford: Blackwell.

Cuny, F. (1983). *Disasters and Development.* Oxford: Oxford University Press.

Dalton, D. (1993). *Mahatma Gandhi: Nonviolent Power in Action.* New York: Columbia University Press.

David, R. S. (1996). Re-democratization in the wake of the 1986 People Power Revolution: Errors and dilemmas. *Kasarinlan, 11,* 5–20.

Delina, L. L. (2016). *Strategies for Rapid Climate Mitigation: Wartime Mobilisation as a Model for Action?* Abingdon: Routledge.

Demski, C., Capstick, S., Pidgeon, N., Sposato, R. G., & Spence, A. (2017). Experience of extreme weather affects climate change mitigation and adaptation responses. *Climatic Change, 140,* 149–164.

Dunlap, R. E., & McCright, A. M. (2011). Organized climate change denial. In J. S. Dryzek, R. B. Norgaard, & D. Schlosberg (Eds.), *The Oxford Handbook of Climate Change and Society* (pp. 144–160). Oxford: Oxford University Press.

Festinger, L. (1962). *A Theory of Cognitive Dissonance.* Stanford: Stanford University Press.

Fisher, D., & McInerney, P.-B. (2012). The limits of networks in social movement retention: On canvassers and their careers. *Mobilization: An International Quarterly, 17*, 109–128.

Fiske, S. T., Cuddy, A. J. C., Glick, P., & Xu, J. (2002). A model of (often mixed) stereotype content: Competence and warmth respectively follow from perceived status and competition. *Journal of Personality and Social Psychology, 82*, 878–902.

Fossil-Free. (2018). *Divestment Commitments.* https://gofossilfree.org/divestment/commitments/.

German Advisory Council on Global Change (WBGU). (2011). *World in Transition: A Social Contract for Sustainability.* Berlin: WBGU.

Gifford, R. (2011). The dragons of inaction: Psychological barriers that limit climate change mitigation and adaptation. *American Psychologist, 66*, 290–302.

Gifford, R., Lacroix, K., & Chen, A. (2018). Understanding responses to climate change: Psychological barriers to mitigation and new theory of behavioral choice. In S. Clayton & C. Manning (Eds.), *Psychology and Climate Change: Human Perceptions, Impacts, and Responses* (pp. 161–183). London: Academic Press.

Haidt, J. (2001). The emotional dog and its rational tail: A social intuitionist approach to moral judgment. *Psychological Review, 108*(4), 814–834.

Haidt, J. (2007). The new synthesis on moral psychology. *Science, 316*, 998–1002.

Harrington, W. (2000). Rosa Parks and the Montgomery Bus Boycott. In P. Winters (Ed.), *The Civil Rights Movement* (pp. 45–57). Cambridge: Cambridge University Press.

Howard, E. (2016, May 7). Philippines investigates Shell and Exxon over climate change. *The Guardian.*

Irfan, U. (2018, October 30). The Supreme Court is about to decide if the children's climate lawsuit can proceed. *Vox.*

Judd, C. M., James-Hawkins, L., Yzerbyt, V., & Kashima, Y. (2005). Fundamental dimensions of social judgment: Understanding the relations between judgments of competence and warmth. *Journal of Personality and Social Psychology, 89*, 899–913.

Kahan, D. M., & Braman, D. (2008). The self-defensive cognition of self-defense. *American Criminal Law Review, 45*, 1–65.

Kahan, D., Jenkins-Smith, H., & Braman, D. (2011). Cultural cognition of scientific consensus. *Journal of Risk Research, 14*, 147–174.

Komisar, L. (1987). *Corazon Aquino: The Story of a Revolution.* New York: George Braziller.

Kraft, P. W., Lodge, M., & Taber, C. S. (2015). Why people "don't trust the evidence": Motivated reasoning and scientific beliefs. *The Annals of the American Academy of Political and Social Science, 658*, 121–133.

Krugman, P. (2018, October 15). Donald and the deadly deniers. *The New York Times.*

Kurzman, C. (2008). Meaning-making in social movements. *Anthropological Quarterly, 81,* 5–15.

Leach, C. W., Ellemers, N., & Barreto, M. (2007). Group virtue: The importance of morality (vs. competence and sociability) in the positive evaluation of in-groups. *Journal of Personality and Social Psychology, 93,* 234–249.

Leiserowitz, A., & Smith, N. (2017). Affective imagery, risk perceptions, and climate change communication. In E. von Storch (Ed.), *Oxford Research Encyclopedia of Climate Science.* Oxford: Oxford University Press.

Lewandowsky, S., & Whitmarsh, L. (2018). Climate communication for biologists: When a picture can tell a thousand words. *PLoS Biology, 16,* e20006004.

Lorenzoni, I., Nicholson-Cole, S., & Whitmarsh, L. (2007). Barriers perceived to engaging with climate change among the UK public and their policy implications. *Global Environmental Change, 17,* 445–459.

Mann, M. E., Hassol, S. J., & Toles, T. (2017, July 12). Doomsday scenarios are as harmful as climate change denial. *The Washington Post.*

Markowitz, E. M., & Shariff, A. F. (2012). Climate change and moral judgement. *Nature Climate Change, 2,* 243–247.

McKibben, B. (2012, July 19). Global warming's terrifying new math. *RollingStone.*

McKibben, B. (2013, April 11). The fossil fuel resistance. *Rollingstone.*

Mooney, C. (2011, May/June). The Science of Why We Don't Believe Science. *Mother Jones.* https://bit.ly/2Ez9egt.

Moser, S. C. (2007). More bad news: The risk of neglecting emotional responses to climate change information. In S. Moser & L. Dilling (Eds.), *Creating a Climate for Change: Communicating Climate Change and Facilitating Social Change* (pp. 64–80). Cambridge: Cambridge University Press.

Moser, S. C. (2010). Communicating climate change: History, challenges, process and future directions. *WIREs Climate Change, 1,* 31–53.

Moyer, B., McAllister, J., Finley, M. L., & Soifer, S. (2001). *Doing Democracy: The MAP Model for Organizing Social Movements.* Gabriola Island: New Society Publishers.

Nepstad, S. E. (2011). *Nonviolent Revolutions: Civil Resistance in the Late 20th Century.* Oxford: Oxford University Press.

Neslen, A. (2015, June 24). Dutch government ordered to cut carbon emissions in landmark ruling. *The Guardian.*

O'Neill, S. J., & Hulme, M. (2009). An iconic approach for representing climate change. *Global Environmental Change, 19,* 402–410.

O'Neill, S. J., & Nicholson-Cole, S. (2009). "Fear won't do it": Promoting positive engagement with climate change through visual and iconic representations. *Science Communications, 30,* 355–379.

Oreskes, N., Conway, E., Karoly, D. J., Gergis, J., Neu, U., & Pfister, C. (2008). The denial of global warming. In S. White, C. Pfister, & F. Mauelshagen (Eds.), *The Palgrave Handbook of Climate History*. London: Palgrave Macmillan.

Parks, R. (1992). *Rosa Parks: My Story*. New York: Dial Books.

Pelling, M., & Dill, K. (2009). Disaster politics: Tipping points for change in the adaptation of socio-political regimes. *Progress in Human Geography, 34*, 21–37.

Roberts, D. (2017, July 11). Did that New York magazine climate story freak you out? Good. *Vox*.

Rosewarne, S., Goodman, J., & Pearse, R. (2014). *Climate Action Upsurge: The Ethnography of Climate Movement Politics*. London: Routledge.

Schock, K. (2005). *Unarmed Insurrections: People Power Movement in Nondemocracies*. Minneapolis: University of Minnesota Press.

Segnit, N., & Ereaut, G. (2007). *Warm Words II: How the Climate Story Is Evolving and the Lessons We Can Learn for Encouraging Public Action*. London, UK: Institute for Public Policy Research.

Sisco, M. R., Bosetti, V., & Weber, E. U. (2017). When do extreme weather events generate attention to climate change? *Climatic Change, 143*, 227–241.

Snow, D. A. (2007). Framing and social movements. In G. Ritzer (Ed.), *The Blackwell Encyclopedia of Sociology*. Malden, MA: Wiley.

Snow, D. A. (2013). Framing and social movements. In D. A. Snow, D. della Porta, B. Klandermans, & D. McAdam (Eds.), *The Wiley-Blackwell Encyclopedia of Social and Political Movements*. Malden, MA: Blackwell.

Snow, D. A., & Benford, R. D. (1988). Ideology, frame resonance, and participant mobilization. *International Social Movement Research, 1*, 197–217.

Spence, A., Poortinga, W., Butler, C., & Pidgeon, N. F. (2011). Perceptions of climate change and willingness to save energy related to flood experience. *Nature Climate Change, 1*, 46–49.

Srnicek, N., & Williams, A. (2016). *Inventing the Future: Postcapitalism and a World Without Work*. London: Verso.

Stern, P. C. (2012). Psychology: Fear and hope in climate messages. *Nature Climate Change, 2*, 572–573.

Stevenson, K., & Peterson, N. (2015). Motivating action through fostering climate change hope and concern and avoiding despair among adolescents. *Sustainability, 8*, 6.

Stirling, A. (2014). Transforming power: Social science and the politics of energy choices. *Energy Research & Social Science, 1*, 83–95.

Stoknes, P. E. (2014). Rethinking climate communications and the "psychological climate paradox". *Energy Research & Social Science, 1*, 161–170.

Stoll-Kleeman, S., O'Riordan, T., & Jaeger, C. C. (2001). The psychology of denial concerning climate mitigation measures: Evidence from Swiss focus groups. *Global Environmental Change, 11*, 107–117.

Swim, J. K., Stern, P. C., Doherty, T. J., Clayton, S., Reser, J. P., Weber, E. U., et al. (2011). Psychology's contributions to understanding and addressing global climate change. *American Psychology, 66,* 241–250.

The Next Systems Project. (2018). https://thenextsystem.org.

Thomas, E. F., McGarty, C., & Mavor, K. I. (2009). Transforming "apathy into movement": The role of prosocial emotions in motivating action for social change. *Personality and Social Psychology Review, 13,* 310–333.

Thompson, M. R. (1995). *The Anti-Marcos Struggle: Personalistic Rule and Democratic Transition in the Philippines.* New Haven: Yale University Press.

Wallace-Wells, D. (2017, July 9). The uninhabitable earth. *New York Magazine.*

Weber, E. U. (2006). Experience-based and description-based perceptions of long-term risk: Why global warming does not scare us (yet). *Climatic Change, 77,* 103–120.

Weber, E. U. (2010). What shapes perceptions of climate change? *WIREs Climate Change, 1,* 332–342.

Zilk, E. (2016). *The Climate Mobilization Victory Plan.* The Climate Mobilization.

CHAPTER 5

Triggering Communal Peer Pressure: Spreading a Shared Understanding of Demands

Abstract Communal peer pressure catalyzes a sense of obligation among social groups, and encourages and nudges people who, at first, may not identify with those groups to blend, fit in, and conform to the larger expectations of those groups. Social comparison—communal peer pressure, horizontal diffusion of climate actions, and instilling a strong sense of common obligation—could be summoned as a motivator for climate actions. Mobilization strategies built around it can complement established drivers of participation such as care for sustainability, forethinking about the future generations of humans and non-humans, and economic incentives.

Keywords Peer pressure · Relationship · Obligation · Unity · Trust · Nudge · In-group mobilization · Common enemy

Convincing people, groups, organizations, and institutions to participate and willingly engage in social actions (even when they do not necessarily *want* to participate)—such as by being physically present in protests, supporting mobilization efforts through resource provision, and prefiguring better futures in their own spheres of influence (e.g. households and communities)—is a challenge for every social movement, yet an important process in mobilization. Organizers would rely upon one persuasion mechanism to achieve this stage: **communal peer pressure**. Over the years, this mechanism has been remarkably effective and proved useful in mobilization. This form of pressure catalyzes a sense of obligation

© The Author(s) 2019   71
L. L. Delina, *Emancipatory Climate Actions*,
https://doi.org/10.1007/978-3-030-17372-2_5

among social groups, such as those found in families, neighbors, church groups, professional clubs, senior clubs, clans, tribes, etc. Communal peer pressure, in time, encourages and nudges people who, at first, may not identify with those groups to blend, fit in, and conform to the larger expectations of those groups (Clark et al. 1987). Communal peer pressure is challenging to describe, because it varies in expression and form from person to person. Something in common, however, is present in this social phenomenon: *how* it is spread.

Psychologists find that communal peer pressure occurs via community ties and expectations (*internal*) (Cialdini et al. 1990; Kallgren et al. 2000; Cialdini 2003; Cialdini and Goldstein 2004) and through mirrored expectations (*external*). There is a natural instinct embedded in social ties—a sympathy that makes people willing to speak, even fight, for someone they "like" when they see or feel that those persons are unjustly treated (Small and Simonsohn 2008). Studies show that people seem to care less about injuries inflicted upon "strangers," but when a "friend" is insulted, a sense of umbrage tends to develop (e.g. Small and Simonsohn 2008).

Dan Kahan, a psychologist and a risk perception expert, and his colleagues, show how ties in and expectations from a community—to which an individual is culturally and socially committed (directly or indirectly, internally or externally)—influence and focus the definition and guidance of that individual's perception of risk (Kahan et al. 2012; cf. Weber 2010). Kahan's studies show that if an individual ignores his or her social obligations to his or her respective communities, and if that individual opts to veer away from their communities' expected patterns, the individual risks losing his or her social standing in the community, including the possibility of being ostracized, snubbed, or excommunicated.

This chapter describes how the four social movements under study counted on and tapped into communal peer pressure in their mobilization strategies. The chapter then describes how the climate action movement has been using communal peer pressure to mobilize public support. The chapter closes with a section on how peer pressure can be effectively triggered by forging relationships across (internal and external) communities.

PEER PRESSURE IN YESTERDAY'S SOCIAL MOVEMENTS

Convincing people to pursue a similar goal, especially when that pursuit entails real hardships—such as the possibility of going to jail for breaking the salt law as in Gandhi's Dandi March, walking to work rather than taking the bus as in the case of Montgomery, going to jail for partici-pating in an unapproved public gathering as in the cases of Manila and Rangoon—is a huge challenge. Community expectations and social peer pressure "to help someone who is close and who is in distress" have helped explain how these campaigns expanded and scaled up from micro-mobilizations in groups of close friends sharing similar identities into macro-mobilizations where communities of varied identities gath-ered together to pursue common objectives.

Gandhi's campaign is instructive with regards to the power of tapping people's relationships with their social groups and communities in mobi-lization. For instance, Gandhi successfully tore down the barriers separat-ing Indians by associating publicly with Muslims and untouchables and by mingling with the poor—not just with his fellow Hindus and elites. He would wear Indian hand-spun cloth (khadi), organize associations to encourage spinning, and making time for it in his daily routine even at times when he was actively leading nationwide campaigns (Brown 1989: 203–204, 206). Gandhi also founded communities where people would live according to the concept of swaraj or "self-rule" to prefigure an independent India.

The Sabarmati Ashram in Ahmedabad in the state of Gujarat is one of the most famous of these communities. In this ashram, members vowed to adhere to truth and *ahimsa* (celibacy), lead simple lives focused on prayer and manual labor, and disregard untouchability (Brown 1989: 100, 106, 109). Gandhi's *swaraj* would offer everyone in India—regardless of religion, caste, and status in life—a community to belong to and an opportunity to have a tangible and meaningful contribution to the Movement. In many Indian villages, Gandhi recruited volunteers to build schools, improve hygiene, and promote crafts. In the process, he would make long-lasting connections with teachers, merchants, law-yers, etc. whom, in later years, turned into Gandhi's most loyal allies (Dalton 1993: 27–28). In these communities, Gandhi showed how rela-tionship-building can result into social regeneration. To Gandhi, this transformation was not the end in itself but rather a groundwork for nonviolent actions against British rule.

Through communal peer pressure, Gandhi hoped to unify Muslims and Hindus, and include women and untouchables in public life. During his march to Dandi, Gandhi walked with some seventy or so people whom he knew were devoted to his principles of self-denial of ashram life and strict discipline. They shone forth not only as examples to other Indians but to send a message of belongingness. Being drawn from different regions, religious communities, and castes, the marchers effectively served as a metaphor for a heterogeneous Indian nation that can be brought together under a new kind of a binding relationship.

Gandhi effectively tapped on these new relationships and communities to use peer pressure. Each time he would stop in a village on his way to Dandi, Gandhi would address assembled crowds to attack the inhumane British-imposed salt tax (Dalton 1993: 112) and declare that his march was every "poor man's battle" (Dalton 1993: 108). He would then pressure village officers to quit their posts and instruct everyone to personally boycott those officers who refused to resign by, for example, refusing to attend celebrations at an officer's house. Gandhi (1930: 312–313) would also insist on appeals that "must always be to the head and the heart, never to fear of force."

Toward the end of the March, Gandhi sent the message on how innovative his salt *satyagraha* was. It was an easy and simple act of civil disobedience to follow and do in that anyone with a bucket or a pot and access to seawater could make salt. Gandhi's leadership was at its greatest at this moment for his ability to mobilize support for his cause by wedding the Movement's self-rule agenda—a goal originally and exclusively cherished only by a small group of Hindu elites—with down-to-earth demands that would offer millions of Indians a stake in the outcome.

Relationships and communities were also tapped in the Montgomery bus boycott. For many African-American communities, the *Brown v. Board of Education* decision of 1954, which ruled that segregation in public schools was illegal, was, of course, a landmark case elevating the cause for civil rights in the United States. But that decision was still largely abstract, and it was unclear how the ruling would be felt locally and beyond the spheres of public schools. Rosa Parks' arrest, however, ignited something unusual within the African-American community in Montgomery.

Earlier on, the City of Montgomery had jailed Claudette Colvin, Mary Louise Smith, and others for violating the city's bus segregation law. But Parks became the symbol of Montgomery's civil rights campaign largely because she was a deeply respected personality in her community. Her arrest triggered a response from existing social groups where she had strong ties with.

Parks was a member of a dozen of Montgomery's social networks. She was the secretary of the Montgomery chapter of the National Association for the Advancement of Colored People (NAACP). She was a member of a Methodist congregation. She was an overseer of a youth organization in a Lutheran church close to her home. She was a volunteer at a shelter. She was a volunteer in a women's group who knit blankets for a local hospital. She was involved in a number of other social groups (Parks 1992). Parks' association with these social networks across town, which cut across almost all of the city's economic and racial lines, brought her "many close friends" to congregate and develop a united response to her arrest. When Parks' "many close friends" learned about her arrest, Montgomery's NAACP already had a planned response to boycott public buses. The group banked on the arrest to effectively mobilize her "many friends" and their natural inclination "to help a friend in distress."

The Montgomery bus boycott, which begun as a support for a "close friend in distress," turned into a city-wide action that would involve almost all of Montgomery's African-American residents. From Parks' friends spreading the word that they would be boycotting the buses, the Movement spread as the campaign morphed into a sense of *obligation* for African-Americans living in the city (Kohl 2000). Three days after Parks' arrest, African-American pastors explained in their Sunday sermons that every African-American church in Montgomery had agreed to the protest (Kohl 2000). The message was clear: the communities were rallying behind Rosa Parks and that it would be an embarrassment for any African-American parishioner to be seen riding a bus on that Monday. People largely conformed to this communal peer pressure for almost one year.

Community peer pressure was also evident when the Movement that ousted Ferdinand Marcos in the Philippines gathered along Epifanio Delos Santos Avenue (EDSA), a major thoroughfare in Manila, in February 1986. Residents from around EDSA knocked on the doors of their neighbors and started to gather in front of the gates where the mutineers were making camp (Cruz-Del Rosario and Dorsey 2013). Some participants had a clear objective in mind that their participation was tantamount to toppling the dictator, but many joined simply because they were prodded by their family members, their co-workers, and their friends (Montiel 2006).

PEER PRESSURE IN TODAY'S CLIMATE ACTIONS

Climate actions are activities and processes that require a planetary response. Yet, it would seem impossible to effect peer pressure in a "community" that is global in scale that would *obligate* everybody's participation. In addition to climate denial, the psychological distance—both temporal and spatial—in doing climate actions contributes to myopia and inaction, and remains one of the key barriers for doing planet-wide, effective, and coordinated actions in the present (Spence et al. 2012; Weber 2006).

The time period—that is, the years when emissions must peak and decline (e.g. projected at mid-century [2050], end-of-the-century [2100], and beyond)—seems very far into the future. Even the Intergovernmental Panel on Climate Change (IPCC 2018) Special Report on 1.5 °C warming frames climate actions as something that can be done in the "next twelve years." With climate actions calculated and projected over years and decades, public engagement tends to largely depend upon the framing of recent weather events (Donner and McDaniels 2013). In part, extreme weather events have made some strong impressions leading to the Movement to call for urgent climate actions; but these remain largely ineffective—the scale and speed required for effective climate actions remain below par; and the sense of obligation failed to materialize. Climate denialism has been effectively extended into climate actions rendered as "someone else's problem" that "can be dealt with in the future."

The climatic events being described by science are also often distant in geography. Climate effects, which are typically the strongest and most devastating in equatorial regions, the poles, mountains, glaciers, etc., are in places that are spatially far-removed from where extensive and deeper climate actions are supposed to occur—i.e. in industrialized societies responsible for climate change. In cases when extreme weather events do occur in these societies (e.g. the 2018 wildfires in California), the temporal and spatial distance to do global climate actions are still too wide to highlight a sense of obligation among rich societies to act on climate. The public perception of the scale of climate impacts—when they occur—also tends to make many people feel helpless, especially considering that these impacts are already locked in even if we rapidly decline our emissions now.

Another reason that today's climate action messaging also seems to be not working is because effective climate actions require activities and processes that would address the increasing concentrations of invisible gases. Reducing industrially produced greenhouse gases—increasing concentrations of which are the cause of climate change—is not the only invisible aspects of climate actions. Energy use, particularly of electricity—where shifts in generation away from fossil fuel combustion toward renewable sources are regarded as key climate actions—is also an almost imperceptible concept. Few people pay attention to the sources of their electricity.

Many present climate actions are planned, centered, and accomplished around the physical manifestations of the root causes of climate change—that is, in the material infrastructures of the fossil fuel systems. Outward-focused activism has been mainly directed through protests, demonstrations, and lobbying against fossil-fuel based development. Examples of these type of activism are the series of Keystone Pipeline protests in the United States; calls for divestment in superannuation funds, endowments in colleges and universities, government pension funds, and insurance; direct actions against coal-fired power plants and fossil fuel extractions; and lobbying for a price on carbon through legislation. This activism, which targets the worst, obstructionist actors for effective climate action, is being diversified, strengthened, and complemented with parallel actions that focus on the other invisible elements of climate denialism: through prefiguration of various desirable futures.

Prefiguring desirable futures has been in the strategy portfolios and campaign repertoires of many groups in the Movement. Imageries and narratives of 100% renewable energy solutions, for example, have been rising in communities, towns, and cities with full renewable energy transition targets (e.g. The Solutions Project). Schools and universities have also become fertile grounds sowed with prefigurative futures that are largely powered by renewable energy. The Sunrise Movement has emerged as an example of this strand.

Peer Pressure in Tomorrow's Climate Actions

Tomorrow's climate actions need to be widespread, encompassing, and universal. To achieve this, social comparison—communal peer pressure, horizontal diffusion of climate actions, and instilling a strong sense of common obligation—could be summoned as a key motivator. Peer pressure complements already established drivers of participation such as care for sustainability, forethinking about the future generation of humans and non-humans, and economic incentives (Tversky and Kahneman 1986; Brekke and Johansson-Stenman 2008). To be clear, peer pressure is different from nudges in that the latter works in social architecture and design without the target individuals or groups necessarily knowing the nudges

(Thaler and Sunstein 2008), while the former requires individuals and groups to have an explicit understanding of the needed changes in behavior.

Nudging is an approach built from psychology and behavioral economics where small changes in choice architectures could yield large impact on consumption behaviors through shifts from an active choice to a passive choice by default. Nudges have been designed around purchasing decisions that lead people to choose energy efficient options (e.g. by providing life cycle costs in large fonts next to the purchase price of appliances (Kallbekken et al. 2012) and reducing food wastage (e.g. by reducing plate size in restaurants (Kallbekken and Sælen 2013). Several future climate action nudges could be envisaged. For example, a carbon emissions tax could be automatically included when buying airfares, one which would require the purchaser to make a conscious decision to opt out during the purchasing process. Another example is restricting parking in cities while improving public transport with lower fares where people will find it faster and more comfortable to take public transport instead of to drive and find parking.

Individual peer pressure can be mustered in a number of ways but studies have suggested that a focus on mindset shifts in communal or group settings, rather than the individual, develops a collective purpose. Groups can be, for example, collectively pressured to adopt efficient behaviors in their electric consumption when they can directly and knowingly compare their own efforts at reducing their power consumption with their neighbors. Studies have further shown that the most committed groups, which also had the greatest economic savings, were not simply looking at conserving energy and realizing some savings but also to be acknowledged for their actions (Ayres et al. 2012; Alcott 2011; Greenberg 2014; Stern 2014). Future climate actions, thus, can be designed around areas that highlight opportunities for group-to-group comparisons, that are regularly monitored, examined, evaluated and rated, and with exemplary performers acknowledged for their contributions.

Alignment is key. This means climate actions must be linked with the immediate and explicit interests and goals of communities and groups. Alignment involves processes of "bridging" (Snow et al. 1986), which refers to linking two or more ideologically congruent but structurally disconnected frames regarding a particular issue. To achieve bridging, the extent to which future climate actions are framed must be experientially commensurate with the past and present lives of the community or group. Mobilizations in these spaces must be about letting people learn through climate actions how they can turn talk into action within their own cultural milieu, while underlining their sense of obligation to contribute to climate actions.

Face-to-face mobilizations remain essential in the bridging exercise. As with histories of mobilizations, engagement with local and community politics and activities are vital approaches. These involve personalized campaigns that would allow the audience to deeply understand the issue. Framing, through personal stories and narratives, tone of voice, choice of words, and other signals, provides essential contexts for people to engage with climate actions. These approaches also nurture **trust** between the individuals participating in the exchange and can, as shown by the case of Gandhi's mobilization techniques, go a long way toward spreading.

Spreading behavioral change is best leveraged through social networks. The histories of social networks in this book's case studies show the effectiveness of getting a message out via messengers who are spatially and temporally closer to the target groups and communities. For example, pastors and preachers attuned with the climate issue and the need to act on it now are better messengers of climate actions in communities where climate scientists and governments are not particularly welcomed as climate action messengers. There is, however, no guarantee that this would work. Pope Francis call for climate action in *Ladatu Si'*, for example, had backfired with conservatives, who resisted the message, defending their pre-existing beliefs instead (Davis 2016).

Such type of in-group information responds directly to the barrier of psychological distance, including the temporal gaps between the future and the present and the spatial geography between the receiver of the message and the intricacies of the required geographic change. This approach also assists in making the case for urgent climate actions very immediate since it puts the message inside a person's locus of control— where people can act on the basis of their capacity and resources. At the same time, these locally produced climate actions can stimulate powerful peer pressure in their community.

Knotting together local, regional, national and international grievances against a "common" enemy, that is the fossil fuel regime complex, is essential. This means holding together many contemporary social action groups under one obligation: to be in the climate action movement. While this movement may have no *conceptual* unity (in terms of common material interests), it has *nominal* unity (where multiple actions and their actors could effectively cohere with and among one another) (Srnicek and Williams 2016). The mobilization of the climate action movement around actions that contribute to the weakening of the fossil fuel regime complex requires articulating a "common" vision in such a way that a variety of actions, campaigns, and struggles could see their interests being expressed in the Movement.

In this sense, the climate action movement could be conceptualized as a Movement of all Movements reflecting a broad array of interests. It is a labor movement where coalitions can be created to overcome the tensions between the environmental program of decreasing emissions with the economic program of fulfilling, purpose-driven, and meaningful jobs. The Movement is a feminist movement in that it recognizes the invisible labor carried out by women, a gender-responsive labor market, and financial independence among women. The Movement is a #BlackLivesMatter, an anti-racist, and indigenous peoples movement in that racial and minority struggles are salient in climate actions. Many in these populations, for instance, either live close to dirty coal-fired power plants or are directly affected by high unemployment, police and military brutality.

A vision of a future that speaks about the heterogeneity of the composition of the climate action movement is essential. This vision provides a coherent narrative articulating how multiple demands could and would share a common antagonist. Mobilizing the many constituencies of the climate action movement under this shared understanding of demands and a "common enemy" gives consistency without necessarily negating differences. A broad spectrum of society, as shown by the histories of large-scale mobilizations need to be brought together as an active force for change.

REFERENCES

Alcott, H. (2011). Social norms and energy conservation. *Journal of Public Economics, 95,* 1082–1095.
Ayres, I., Raseman, S., & Shih, A. (2012). Evidence from two large field experiments that peer comparison feedback can reduce residential energy usage. *The Journal of Law, Economics & Organization, 29,* 992–1022.
Brekke, K. A., & Johansson-Stenman, O. (2008). The behavioural economics of climate change. *The Oxford Review of Economic Policy, 24,* 280–297.
Brown, J. M. (1989). *Gandhi: Prisoner of Hope.* New Haven: Yale University Press.

Cialdini, R. B. (2003). Crafting normative messages to protect the environment. *Current Directions in Psychological Science, 12,* 105–109.

Cialdini, R. B., & Goldstein, N. J. (2004). Social influence: Compliance and conformity. *Annual Review of Psychology, 55,* 591–621.

Cialdini, R. B., Reno, R. R., & Kallgren, C. A. (1990). A focus theory of normative conduct: Recycling the concept of norms to reduce littering in public places. *Journal of Personality and Social Psychology, 58,* 1015–1026.

Clark, M. S., Ouellette, R., Powell, M. C., & Milberg, S. (1987). Recipient's mood, relationship type, and helping. *Journal of Personality and Social Psychology, 53,* 94–103.

Cruz-Del Rosario, T., & Dorsey, J. M. (2013). Street, shrine, square, and soccer pitch: Comparative protest spaces in Asia and the Middle East. *Air & Space Power Journal-Africa & Francophonie, 4,* 80–96.

Dalton, D. (1993). *Mahatma Gandhi: Nonviolent Power in Action.* New York: Columbia University Press.

Davis, N. (2016, October 24). Pope Francis's edict on climate change has fallen on closed ears, study finds. *The Guardian.* https://bit.ly/2dPExbg.

Donner, S. D., & McDaniels, J. (2013). The influence of national temperature fluctuations on opinions about climate change in the U.S. since 1990. *Climatic Change, 118,* 537–550.

Gandhi, M. K. (1930). *Collected Works of Mahatma Gandhi* (Vol. 43). New Delhi: Digital Library of India.

Greenberg, M. R. (2014). Energy policy and research: The under-appreciation of trust. *Energy Research & Social Science, 1,* 152–160.

IPCC. (2018). Summary for Policymakers. In V. Masson-Delmotte, P. Zhai, H.-O. Pörtner, D. Roberts, J. Skea, P. R. Shukla, A. Pirani, & W. Moufouma-Okia, C. Péan, R. Pidcock, S. Connors, J. B. R. Matthews, Y. Chen, X. Zhou, M. I. Gomis, E. Lonnoy, T. Maycock, M. Tignor, & T. Waterfield (Eds.), *Global Warming of 1.5°C. An IPCC Special Report on the impacts of global warming of 1.5°C above pre-industrial levels and related global greenhouse gas emission pathways, in the context of strengthening the global response to the threat of climate change, sustainable development, and efforts to eradicate poverty* (p. 32). Geneva, Switzerland: World Meteorological Organization.

Kahan, D. M., Peters, E., Wittlin, M., Slovic, P., Ouellette, L. L., Braman, D., & Mandel, G. (2012). The polarizing impact of science literacy and numeracy on perceived climate change risks. *Nature Climate Change, 2,* 732–735.

Kallbekken, S., & Sælen, H. (2013). 'Nudging' hotel guests to reduce food waste as a win–win environmental measure. *Economics Letters, 119,* 325–327.

Kallbekken, S., Sælen, H., & Hermansen, E. A. T. (2012). Bridging the energy efficiency gap: A field experiment on lifetime energy costs and household appliances. *Journal of Consumer Policy, 36,* 1–16.

Kallgren, C. A., Reno, R. R., & Cialdini, R. B. (2000). A focus theory of normative conduct: When norms do and do not affect behavior. *Personality and Social Psychology Bulletin, 26,* 1002–1012.

Kohl, H. (2000). Rosa Parks and the Montgomery bus boycott. In J. Birnbaum & C. Taylor (Eds.), *Civil Rights Since 1787: A Reader on the Black Struggle.* New York: New York University Press.

Montiel, C. J. (2006). Political psychology of nonviolent democratic transitions in Southeast Asia. *Journal of Social Issues, 62,* 173–190.

Parks, R. (1992). *Rosa Parks: My Story.* New York: Dial Books.

Small, D. A., & Simonsohn, U. (2008). Friends of victims: Personal experience and prosocial behaviour. *Journal of Consumer Research, 35,* 532–542.

Snow, D. A., Rochford, E. B., Worden, S. K., & Benford, R. D. (1986). Frame alignment processes, micromobilization, and movement participation. *American Sociological Review, 51,* 464–481.

Spence, A., Poortinga, W., & Pidgeon, N. (2012). The psychological distance of climate change: Psychological distance of climate change. *Risk Analysis, 32,* 957–972.

Srnicek, N., & Williams, A. (2016). *Inventing the Future: Postcapitalism and a World Without Work.* London: Verso.

Stern, P. (2014). Individual and household interactions with energy systems: Toward integrated understanding. *Energy Research & Social Science, 1,* 41–48.

Thaler, R. H., & Sunstein, C. R. (2008). *Nudge: Improving Decisions About Health, Wealth, and Happiness.* New Haven: Yale University Press.

Tversky, A., & Kahneman, D. (1986). Framing of Decisions. *Economic Theory, 251,* 8.

Weber, E. U. (2006). Experience-based and description-based perceptions of long-term risk: Why global warming does not scare us (yet). *Climatic Change, 77,* 103–120.

Weber, E. U. (2010). What shapes perceptions of climate change? *WIREs Climate Change, 1,* 332–342.

CHAPTER 6

Boosting Publicity: Old and New Media, Deliberations, and Organic Ideology Articulation

Abstract Publicizing campaigns need not only be far-reaching but also empowering. Mobilization, thus, requires all forms of traditional and emergent media, at the same time that venues for iterative knowing, probing, and problem-solving and formulating and articulating an organic ideology are created and sustained. With climate denial deeply entrenched in governments and traditional mass media channels, a new hegemony that support emancipatory and transformative climate actions can be promoted using new channels provided by social media; albeit ensuring authentic participation. Strengthening proven modes of face-to-face public engagement remains important. Authentic and influential deliberations can be tapped not only to increase public knowledge but also to open up a culture for reflexive problem solving, which has long-lasting effects to people.

Keywords Publicity · Mass media · Social media ·
Climate communication · Climate denial · Authentic participation ·
Deliberation · Intellectual organization · Organic intellectuals

Getting widespread publicity is key to ensure a successful Movement (Moyer et al. 2001). Mass media promotions, including those through social media (Bennett and Sagerberg 2012), are the most cost-effective ways of publicizing campaigns largely because these platforms are far-reaching and cheaper than personal interactions. Since

© The Author(s) 2019
L. L. Delina, *Emancipatory Climate Actions*,
https://doi.org/10.1007/978-3-030-17372-2_6

participation in social actions is gained through personal connections (which are important as described in Chapter 5), as well as through other channels (such as mass media, social media, websites, and mailing lists) (Klandermans et al. 2014; Walgrave and Klandermans 2010), boosting publicity requires tapping all possible communication channels. Widespread conversations around social issues are a signal that the Movement is at its peak; hence getting to this stage is key.

Climate deniers, including fossil fuel actors and their lobbyists, and their supporters in governments, consistently undermine climate science and the need to act on it quickly, decisively, and effectively by amplifying the level of uncertainty about the relationship between climate change and human activity (Oreskes and Conway 2012). The media exaggerates this situation further with "false balance" where deniers are afforded the opportunity to advance their untruthful claims in media platforms (Boykoff 2011). In climate communication, therefore, fair, honest, and truthful information and its timely conveyance to the public, using tools that assist (rather than mar and preclude) in the easy and better understanding of the issue is a requirement for closing the disparities between public opinion and scientific facts.

This chapter describes how the histories under study publicized the truths of their respective campaigns using both traditional and innovative communication channels. The chapter then describes how the climate action movement uses multiple communication channels, highlighting the role of social media in their contemporary campaigns. The chapter closes with a section on how future climate actions can be publicized by going back to the basics of face-to-face conversations.

PUBLICITY IN YESTERDAY'S SOCIAL MOVEMENTS

Gandhi performed his march from his ashram in Sabharmati to the seaside town of Dandi as a political theater, playing vividly not only to onlookers, but also through the media, to the entire subcontinent and to the world (Brown 1989: 97). The 400-kilometer march was cleverly designed such that Gandhi and the marchers would make stops in villages where they could do constructive campaigning. To publicize the march, the schedule of the stops was published in *Navajivan*, a Gujarati-language weekly. Along the way, Gandhi wrote articles and gave interviews. Indian newspapers carried stories of the march on their front pages. Three Bombay-based cinema companies sent their crews along to shoot footages of the march. Foreign journalists were also present and, through their accounts, turned Gandhi into a household name in Europe and America. By end of 1930, Gandhi was named *Time* magazine's "Man of the Year" (Brown 1977: 104–105).

Extensive media coverage of Gandhi's campaigns also included the violent dispersal of nonviolent marchers in Dharasana, which occurred a few months after the salt-making campaign in Dandi. Such coverage of the dispersal proved very useful to mobilize the Movement. American war journalist and Pulitzer Prize winner Webb Miller (in Weber 1997: 443–447), who covered the Dharasana campaign for *The United Press*, reported on the violent dispersal as follows:

> Suddenly, at a word of command, scores of native police rushed upon the advancing marchers and rained blows on their heads with their steel-shod *lathis* [clubs]. Not one of the marchers raised an arm to fend off the blows. They went down like ten-pins...[with] sickening whacks of the clubs on unprotected skulls.... In two or three minutes the ground was quilted with bodies. Great patches of blood widened on their white clothes. The survivors without breaking ranks silently and doggedly marched on until struck down.... the spectacle of unresisting men being methodically bashed into a bloody pulp sickened me.... I felt an indefinable sense of helpless rage and loathing, almost as much against the men who were submitting unresistingly to being beaten as against the police wielding the clubs.

Miller's story helped launch the Movement into a theater of dramatic proportions both for the Indian public and international observers to behold. As Miller published his account in 1350 newspapers worldwide, millions were able to read about the selfless suffering of the Indians. The story reached the US Senate where it was read and its text officially admitted into the minutes (Singhal 2010). As publicity spread, Miller's report helped sway moral righteousness from the side of imperial Britain to the cause of Free India. More importantly, however, was the trigger that Miller's reporting provided in terms of more peaceful demonstrations throughout India. This was a clear demonstration of "backfire," where the violent dispersals of peaceful protesters, instead of them being corralled, created more support and attention for the protesters (cf. Martin 2007).

The American civil rights movement, during the Montgomery bus boycott also attracted significant media attention. Journalists from around the United States and the world converged in Montgomery and made Rosa Parks and Martin Luther King, Jr. household names (Fairclough 1987: 27–29).

The Philippine case study illustrates how powerholders can exercise strong control over media entities who used them for official propaganda. The absence of a free press in Marcos' Philippines and the dearth of independent media during the culmination of the generally peaceful 1986 People Power Revolution prompted citizen-initiated public broadcasts to publicize this pivotal event (Gonzales 1988). During Marcos' martial law years, the Philippine Catholic Church-owned *Radio Veritas* was the only independent radio station. *Radio Veritas* was alone in broadcasting the news about Ninoy Aquino's assassination and the subsequent government investigation. The radio station was also where Jaime Cardinal Sin, Archbishop of Manila, went on air to ask people to support the mutineers along the police and military camps at the Epifanio Delos Santos Avenue (EDSA), Manila's major thoroughfare and site of the Revolution. Government troops, however, knocked down the radio station's transmitter hours after the Cardinal's appeal.

Contrary to the government's assumption that they were able to suppress the broadcast, *Radio Veritas* broadcaster June Keithley was still able to go on air through a newly set-up, secular, clandestine radio station they called *Radyo Bandido* (translation: Outlaw Radio) (Horsfield 1995). At noon of the first day of the People Power Revolution, a few hours after the knocking down of *Radio Veritas'* transmitter, *Radyo Bandido* became a reliable and effective source of news and information for many Filipinos. Keithley provided news updates and encouraged the protesters in EDSA. Through *Radyo Bandido's* broadcasts, which reached provinces outside Manila, thousands more were mobilized (Cruz-Del Rosario and Dorsey 2013). Provincial radio stations then hooked onto *Radyo Bandido* to deliver a blow-by-blow account of the ongoing Revolution in the capital. Inflamed by these broadcasts, people from other islands were poised to come by boat to Manila, pre-empted only by later radio broadcasts that Marcos already left for Hawai'i (Enriquez 2006).

PUBLICITY IN TODAY'S CLIMATE ACTIONS

As shown by the histories under study, getting widespread publicity is a factor in ensuring successful mobilizations (Moyer et al. 2001). Not all mobilizations, however, were granted identical or appropriate media exposure and coverage. The contemporary climate action movement has been presented with this dilemma.

Low media attention and unfriendly media stance over climate issues have been prominent in high emission countries such as Australia and the United States (Bacon 2013; Bacon and Nash 2012; Feldman et al. 2013). In these places and elsewhere, climate change reporting remains narrowly framed either as an impending or an emerging environmental problem with serious consequences in the future instead of it being an ongoing and present social, economic, and political issue (Boykoff and Boykoff 2004) requiring effective climate actions right now. Even positive stories about climate actions, such as sustainable energy transitions or ambitious renewable energy targets, often do not merit as much airtime compared to stories on violence, wars, and political bickering. As an example, when the International Panel on Climate Change (IPCC) released its 1.5 °C special report in Seoul, mainstream US-based mass media hardly reported on it.

When climate change is reported in the news, many mainstream media outlets, particularly in the United States and Australia, have instead shown bias against climate actions (Feldman et al. 2013; Bacon and Nash 2012). In interviewing climate scientists, for example, equal time is often afforded to non-scientist deniers of climate science. Rupert Murdoch's *Fox News*, the *Wall Street Journal*, and the *New York Post* have, for years, been providing opportunities for representatives of conservative think tanks, climate science deniers, and contrarian scientists to consistently ridicule and assault climate change, the IPCC, and climate scientists (Dunlap and McCright 2011).

The Australian Centre for Independent Journalism, in its report on the coverage of climate change in Australian print publications, also demonstrates the persistent misleading and confusing news reports about scientific findings on climate science, especially in Murdoch-owned presses (Bacon 2013). Misinformation and doubt-sowing reporting in Australia includes *The Australian*, the country's only national newspaper for a general audience (Bacon 2013). Murdoch's *The Daily Telegraph*, not surprisingly, has subjected my own work on rapid climate mitigation, alongside those by other Australian-based climate scientists, to a pejorative attack (see Thomas 2013).

The fossil fuel regime, which has extended its reach and influence in mainstream media, has been impacting the framing of climate actions, as demonstrated for instance by the powerful Australian coal industry

(Bacon and Nash 2012). Rupert Murdoch's *News Corporation* through its tabloid *The Daily Telegraph* in Sydney and *The Courier Mail* in Brisbane continue to carry news stories based on the primary assumption that the growth and expansion of the coal industry is favorable and necessary for Australia (Bacon and Nash 2012) and, therefore, must be protected. The attitude of Australian media toward climate change remains, at best, tilted toward protecting the country's coal industry.

The absence of media coverage of the climate action movement by mainstream broadcasters was also evident during the 2014 People's Climate March. Despite the March's huge turnout in the streets of New York city, mainstream news outlets failed to cover the campaign (Johnson 2014). Only *Democracy Now!* an independent media organization, broadcasted the event live in its entirety. *NBC Nightly News* was the sole primetime news program in major US TV channels to air a segment about the March; ABC used approximately 23 seconds on the topic in its own newscast; CBS did not mention it at all (Mirkinson 2014).

The Movement is now relying heavily on the Internet, particularly on social media, to provide inexpensive publicity outlets to distribute its messages and mobilize support. Among the many benefits of this platform is the emergence of national and international climate action networks that allowed for large-scale mobilization , while, at the same time, providing a more decentralized, self-replicating, and even self-correcting alternative mobilization platform (Orr 2013). Social media have become a foremost tool in modern mobilization as evidenced for instance by an approximately 630,000 social media posts generated during the 2014 People's Climate March (People's Climate March 2014) and the extensive use of Twitter during the Keystone XL pipeline protests (Hodges and Stocking 2016). The use of social media in mobilization for climate has also extended beyond campaigning to include the documentation and coordination of prefigurative climate actions such as through sharing best practices for energy efficiency and access to renewable energy solutions.

PUBLICITY IN TOMORROW'S CLIMATE ACTIONS

The public at large needs to comprehend the threats of climate change and its holistic solutions; otherwise, they cannot be expected to endorse, support, and engage with climate actions. Online and social media may help boost the audience reach—especially given the lackluster treatment of climate science and climate actions by mainstream media outlets. This is particularly true since Internet-based news sources have already charted their own virtual territories, claiming audiences traditionally captured by traditional media institutions. Indeed, mainstream media agencies, already grappling with lost TV viewership, radio listenership, and magazine and newspaper readership (Orr 2013; Slaughter 2005), have already started transitioning to digital.

The climate action movement needs to expand its media strategy to take advantage of all available media platforms. With audiences having varied preferences in terms of the kind of climate actions they could and wanted to pursue, tapping this variety of available sources in communicating effective climate actions make sense (Agyeman et al. 2007; Featherstone et al. 2009). In the absence of support from mainstream media organizations, the Movement may find inspiration from the Philippine case, where citizen-initiated radio broadcasts helped to publicize and mobilize the 1986 People Power Revolution (Gonzales 1988). Progressive media channels such as *Democracy Now!* should not shy away from communicating the message of doing climate actions, while, at the same time, producing materials that are not only informative but also more accessible and entertaining.

Regardless of mainstream media entities' treatment of the climate issue and the required large-scale climate actions, the Movement still needs to convince the mainstream media into reporting and accepting climate actions' key ideas and messages. Engaging mainstream media remains essential, especially because the digital revolution has failed to provide everybody an audience (despite it enabling people who have access to it to have a voice). Traditional media channels—TV and radio networks, daily newspapers, and magazines—have remained indispensable in mobilization. With the ability of these types of media to influence and alter public opinion remaining strong, the Movement needs to recognize that it will need to alter the ways it presents climate actions as a holistic system change project.

Despite its trumpeted benefits of 24-hour access to information and promise of leveling the playing fields, the Internet has, indeed, brought with it profound and complex social issues. These issues could easily impinge on the larger purposes of the climate action movement: that of building a counter-hegemony built on concepts of justice and sustainability. It is imperative, therefore, that the Movement devises ways using Internet-based technologies to secure and ensure the authenticity of members who participate through and in these platforms.

By authenticity, I mean the degree to which people would engage themselves in actual and reasoned discourses on what would constitute effective climate actions, rather than the mere strategic pursuit of interest or symbolic participation. For example, electronic action or "clicktivism," such as through pushing a button to add one's name to an e-petition, sending an email to a member of Parliament, or "liking" or "following" a climate action group on social media are largely unthinking, isolated, and one-way (cf. Schlosberg and Dryzek 2002). These approaches—although they may lead to something useful—has, instead, depleted social capital used to enliven energetic social actions in the past.

The rise of untrustworthy information, or fake news, in the Internet era also exposes the many vulnerabilities of society. The preponderance of fake news, purveyed primarily and blasted by paid troll farms, in social media, has become one of contemporary society's major challenges. There is no panacea to the problem of fake news but reflexivity remains key. In this regard, the Movement must exert efforts to also question how technology itself not only helps construct or reconstruct but also deconstruct and destroy human societies. Designing strategies that ensure only credible and factual information is relayed to the public by the Movement remains important.

Given the dearth of coverage from mainstream media and the quirks of social media, mobilization for effective climate actions will still entail face-to-face engagements. Future campaigns, therefore, must deal with ways to strengthen the quality of presentations, seminars, and conversations in homes, schools, cafeterias, businesses, community fora, and places where large groups of people converge, including houses of worship. Similar to neighbor-to-neighbor mobilizations for Free India, civil rights in the United States, and church-based mobilizations in 1980s Philippines, face-to-face climate action conversations should stress the importance of a mobilized citizenry as an active political force for emancipatory and transformative change.

Future mobilization meetings can be focused not only on educating and training people on climate science and available climate actions, but also, and most important, on empowering people to turn climate talk into climate actions. The focus of mobilization has to deliberately shift toward achievable climate actions. As histories under study suggest, Movements only become more effective when campaigns self-propel.

This stage is reached only when people start to feel ownership of the action itself. This process requires people to engage with the action—an end which can be prodded using platforms for inclusive, authentic, and influential deliberations.

Deliberations have already been well documented in many social change domains (Lang 2007; Girard et al. 2003; Dryzek et al. 2009; Niemeyer et al. 2011). These exercises are also present in sustaining community-based climate actions. In the indigenous Sami populations in Finnmark, in northeastern Norway, for example, deliberation proved useful when deciding on the location and speed of a wind farm construction (McCauley et al. 2016). Deliberations also worked in the context of a rural, community-oriented sustainable energy transition in Thailand (Delina 2018).

Deliberations would involve a series of professionally facilitated meetings where people can deliberate with one another in a respectful environment to learn and locate possible climate actions that they can take, support, or lobby for (Dryzek 2000). These exercises are processes of iterative knowing, probing, and solving that consider the variety of opinions and suggestions presented and supported by a heterogeneous mix of citizen-participants. Unlike more formal deliberative avenues such as parliaments, the final output of these exercises would not necessarily be a consensus, but a set or a list of climate actions generated by the participants themselves (Dryzek 2000; Carson 2011). Over time, pockets of deliberations on climate actions can be linked into a larger deliberative system, hence contributing to the process of webbing (Delina 2018; Niemeyer and Jennstål 2018; Mansbridge et al. 2012; also see Chapter 7).

With the Movement advancing and mobilizing for holistic, system-change, emancipatory, and transformative agenda for a new hegemony (see Chapter 3), it is key to ensure that its strategies will also involve intellectual organizations such as think tanks, colleges, and universities. The Climate Leadership Network, comprised of colleges and universities across all fifty US states including the District of Columbia, have committed to act on climate and prepare their students through research and education to solve the climate change challenge. Twelve visionary college and university presidents had initiated the American College & University Presidents' Climate Commitment in 2006, which became the precursor to this 600 US colleges and universities-strong Climate Leadership Network. While this effort is commendatory, commitments for climate actions in higher education and research, nevertheless, have to include the actual development of "organic intellectuals," whom Antonio Gramsci conceptualizes as social agents formulating and articulating an organic ideology (Ramos 1982), to develop long-term proposals for meeting the vision of highly sustainable economies. To gain its full effect, these emerging proposals have to be fed back into the narratives of climate actions, where they will be delivered free of jargons and resonate with everyday conversations.

REFERENCES

Agyeman, J., Doppelt, B., Lynn, K., & Hatic, H. (2007). The climate-justice link: Communicating risk with low-income and minority audiences. In S. C. Moser & L. Dilling (Eds.), *Creating a Climate for Change: Communicating Climate Change and Facilitating Social Change* (pp. 119–138). Cambridge: Cambridge University Press.

Bacon, W. (2013). *Climate Science in Australian Newspapers*. Sydney: The Australian Centre for Independent Journalism.

Bacon, W., & Nash, C. (2012). Playing the media game: The relative (in)visibility of coal industry interests in media reporting of coal as a climate change issue in Australia. *Journalism Studies, 13*, 243–258.

Bennett, W. L., & Sagerberg, A. (2012). The logic of connective action: Digital media and the personalization of contentious politics. *Information, Communication & Society, 15*, 37–41.

Boykoff, M. T. (2011). *Who Speaks for the Climate? Making Sense of Media Reporting on Climate Change*. Cambridge: Cambridge University Press.

Boykoff, M. T., & Boykoff, J. M. (2004). Balance as bias: Global warming and the US prestige press. *Global Environmental Change, 14*, 125–136.

Brown, J. M. (1977). *Gandhi and Civil Disobedience*. Cambridge: Cambridge University Press.

Brown, J. M. (1989). *Gandhi: Prisoner of Hope*. New Haven: Yale University Press.

Carson, L. (2011). Dilemmas, disasters and deliberative democracy: Getting the public back into policy. *Griffith Reviews, 32*, 25–32.

Cruz-Del Rosario, T., & Dorsey, J. M. (2013). Street, shrine, square, and soccer pitch: Comparative protest spaces in Asia and the Middle East. *Air & Space Power Journal - Africa & Francophonie, 4*, 80–96.

Delina, L. (2018). Climate mobilizations and democracy: The promise of scaling community energy transitions in a deliberative system. *Journal of Environmental Policy & Planning*. https://doi.org/10.1080/15239 08x.2018.1525287.

Dryzek, J. (2000). *Deliberative Democracy and Beyond: Liberals, Critics, Contestations*. Oxford: Oxford University Press.

Dryzek, J., Belgiorno-Nettis, L., Carson, L., Hartz-Karp, J., Lubensky, R., Marsh, I., et al. (2009). The Australian citizens' parliament: A world first. *Journal of Public Deliberation, 5,* 1–7.

Dunlap, R., & McCright, A. (2011). Organized climate change denial. In J. Dryzek & R. Schlosberg (Eds.), *The Oxford Handbook of Climate Change and Society* (pp. 144–160). Oxford: Oxford University Press.

Enriquez, E. L. (2006). Media as site of social struggle: The role of Philippine radio and television in the EDSA Revolt of 1986. *Plaridel, 3,* 53–82.

Fairclough, A. (1987). *To Redeem the Soul of America: The Southern Christian Leadership Conference and Martin Luther King, Jr.* Athens: University of Georgia Press.

Featherstone, H., Weitkamp, E., Ling, K., & Burnett, F. (2009). Defining issue-based publics for public engagement: Climate change as a case study. *Public Understanding of Science, 18,* 94–101.

Feldman, L., Maibach, E. W., Roser-Renouf, C., & Leiserowitz, A. (2013). Climate on cable: The nature and impact of global warming coverage on Fox News, CNN, and MSNBC. *The International Journal of Press/Politics, 17,* 3–32.

Girard, M., Poletta, F., & Stark, D. (2003). *Policy Made Public: Technologies of Deliberation and Representation in Rebuilding Lower Manhattan*. New York: Center on Organizational Innovation, Columbia University.

Gonzales, H. (1988). Mass media and the spiral of silence: The Philippines from Marcos to Aquino. *Journal of Communication, 38,* 33–48.

Hodges, H. E., & Stocking, G. (2016). A pipeline of tweets: Environmental movements' use of Twitter in response to the Keystone XL pipeline. *Environmental Politics, 25,* 223–247.

Horsfield, B. (1995). Communication for social revolution. *Journal of International Communication, 2,* 52–66.

Johnson, T. (2014, September 21). Sunday news shows ignore historic climate march. *Media Matters for America*.

Klandermans, B., van Stekelenburg, J., Damen, M. L., et al. (2014). Mobilization without organization: The case of unaffiliated demonstrators. *European Sociological Review, 30,* 702–716.

Lang, A. (2007). But is it for real? The British Columbia Citizens' Assembly as a model of state-sponsored-citizen empowerment. *Politics & Society, 35,* 35–69.

Mansbridge, J., Bohman, J., Chambers, S., Christiano, T., Fung, A., Parkinson, J., et al. (2012). A systemic approach to deliberative democracy. In J. Parkinson & J. Mansbridge (Eds.), *Deliberative systems: Deliberative Democracy at the Large Scale* (pp. 1–26). Cambridge: Cambridge University Press.

Martin, B. (2007). *Justice Ignited: The Dynamics of Backfire.* Lanham: Rowman & Littlefield.

McCauley, D., Heffron, R., Pavlenko, M., Rehner, R., & Holmes, R. (2016). Energy justice in the Arctic: Implications for energy infrastructural development in the Arctic. *Energy Research & Social Science, 16,* 141–146.

Mirkinson, J. (2014, September 22). TV news misses yet another opportunity to cover climate change. *The Huffington Post.*

Moyer, B., McAllister, J., Finley, M. L., & Soifer, S. (2001). *Doing Democracy: The MAP Model for Organizing Social Movements.* Gabriola Island: New Society Publishers.

Niemeyer, S., & Jennstål, J. (2018). Scaling up deliberative effects: Applying lessons of mini-publics. In A. Bächtinger, J. Dryzek, J. Mansbridge, & M. E. Warren (Eds.), *The Oxford Handbook of Deliberative Democracy* (pp. 329–347). Oxford: Oxford University Press.

Niemeyer, S., Hobson, K., Russel, J., Ord-Evans, I., Boswell, J., & dos Santos, E. (2011). *Participant Recommendations and Report: Climate Change and the Public Sphere Project* (Centre for Deliberative Democracy & Global Governance Working Paper 2011/4). Canberra, Australia: Centre for Deliberative Democracy & Global Governance, Australian National University.

Oreskes, N., & Conway, E. M. (2012). *Merchants of Doubt: How a Handful of Scientists Obscured the Truth on Issues from Tobacco Smoke to Global Warming.* New York: Bloomsbury Press.

Orr, D. W. (2013). Governance in the long emergency. In E. Assadourian & T. Prugh (Eds.), *State of the World 2013: Is Sustainability Still Possible?* (pp. 279–291). Washington, DC: Island Press.

People's Climate March. (2014). *Wrap Up.* http://bit.ly/1qxNEZt.

Ramos, V., Jr. (1982). The concepts of ideology, hegemony, and organic intellectuals in Gramsci's Marxism. *Theoretical Review, 27.* https://www.marxists.org/history/erol/periodicals/theoretical-review/1982301.htm.

Schlosberg, D., & Dryzek, J. S. (2002). Digital democracy: Authentic or virtual? *Organization & Environment, 15,* 332–335.

Singhal, A. (2010). The Mahatma's message: Gandhi's contributions to the art and science of communication. *China Media Research, 6,* 103–106.

Slaughter, A.-M. (2005). *A New World Order.* Princeton: Princeton University Press.

Thomas, T. (2013, December 19). A room full of eco-idiots. *The Daily Telegraph.*

Walgrave, S., & Klandermans, B. (2010). Open and closed mobilization patterns: the role of channels and ties. In S. Walgrave & D. Rucht (Eds.), *The World Says No to War: Demonstrations Against the War on Iraq* (pp. 169–193). Minneapolis: University of Minnesota Press.

Weber, T. (1997). *On the Salt March*. New Delhi: HarperCollins.

CHAPTER 7

Diversifying Networks: Webbing Heterogeneous Actors and Their Plural Campaigns

Abstract In mobilization, participant and campaign diversity is as important as webbing multiple, yet fragmented, actions. A networked approach to mobilization, however, does not mean centralizing actions through top-down institutional arrangements; rather, it means creating healthy, sustaining ecosystem of polycentric structures where power emanates from multiple centers of influence. Under an overarching vision of a common good, climate actions and their actors would interact, interconnect, and cohere with each other in this ecosystem-while acknowledging that this will be messy and contested processes. As many climate action participants and their respective campaigns interconnect their diverse sources of power, capacities and practices—while remaining respectful of each other's contributions, expertise, and experience—the climate action movement projects itself as a Movement of all Movements.

Keywords Networks · Webbing · Diversity · Heterogeneity · Institutional arrangement · Polycentricity · Pluralism · Cosmopolitanism

Individual actions may seem tiny, but because of webbing multiple, varied, and heterogeneous yet similarly oriented groups, they can multiply and, thus, quickly achieve scale. As a Movement grows in number, the diversity of campaigns and its participants also needs to expand. Although individuals and groups may carry out different forms of social movement activities at different times, under different organizations,

© The Author(s) 2019
L. L. Delina, *Emancipatory Climate Actions*,
https://doi.org/10.1007/978-3-030-17372-2_7

using different tactics, and focused on different audiences, they should still work toward the same ultimate end (Schock 2005).

The more diverse the participants are—in terms of gender, age, religion, ethnicity, ideology, profession, socioeconomic status, etc.—the better the chances of success, according to histories of mobilization. With diversity, success tends to be higher since adversaries can find it more difficult to isolate mobilization participants (cf. Chenoweth and Stephan 2011). The likelihood of tactical and strategic diversity is also increased with different groups bringing in different forms of campaigns and their own unique capacities to the Movement (Chenoweth and Stephan 2011; Murphy 2005; Van Dyke and McCammon 2010).

A Movement is, indeed, best described as constellations of heterogeneous actors and groups with differing emphases on the nature of the challenge but all contributing to the larger cause using their own unique capacities. Because of this arrangement, Movements generally have informal institutional structures; that is, an absent hierarchy and no central authority, hence posing the challenge of—and the need for—webbing (Andrews et al. 2010; Della Porta and Diani 2006). Coalitions and networks—the end states of the processes of webbing—define the Movement's unique "structure" (Rhodes 1996). These interconnections offer advantages for mobilization in that webs help portray a Movement as a visually huge, systematic assemblage. The impression of being "big" causes greater public impact, makes the movement a stronger lobbying force, and allows groups to combine their resources (Murphy 2005).

This chapter describes how the histories under study webbed multiple groups that act on their respective issues together in networks, coalitions, and alliances; and how these interconnections strengthened their campaigns and their movements at large. The chapter then describes how climate action groups, individuals, and institutions are currently webbed and closes with a section on how webbing of future climate actions can be effectively incorporated into a polycentric system.

WEBBING IN YESTERDAY'S SOCIAL MOVEMENTS

Gandhi knew that he and his associates could not direct the Movement for Free India in every Indian city, town, or village, especially once they were arrested for their nonviolent protests (an eventuality they aptly prepared for); hence, they left the mobilizing efforts up to a webbed system of provincial committees, which devised local campaign tactics that would take advantage of local, context-specific conditions. These decentralized actions, however, came with one proviso: they had to be strictly nonviolent (Gandhi 1930: 135–137).

In non-coastal areas where a salt *satyagraha* could not be obviously performed, people participated in cloth boycotts and other forms of civil disobedience such as the refusal to pay rent. Other campaigners forced their local officials to choose between repression and retreat. Indian women also participated ardently, with many even becoming leaders of provincial civil disobedience campaigns. It is key to note, however, that the Movement's failure to bring in Hindus, Muslims, and Sikhs together meant that it did not really speak for "all" of India. This proved a big weakness as it foreshadowed the latter division of the sub-continent, as well as the ensuing bitter communal strife across India that far outlasted the British.

The Philippine case also illustrates the strength of a networked approach to mobilization. In 1985, *Bagong Alyansang Makabayan* (BAYAN; translation: New Nationalist Alliance) provided the broader umbrella under which several progressive and social action groups could ally themselves under. BAYAN was formed two years after Ninoy Aquino's assassination and a year before the snap national elections. The Philippine Catholic Church had also been organizing grassroots networks through small, personalized, and face-to-face group meetings in basic Christian communities in its parishes nationwide.

BAYAN, together with Church-supported networks, also organized large-scale anti-Marcos campaigns. These included: a massive strike throughout Mindanao, Philippines second largest island; a nine-day sit-in in the front of the offices of the Ministry of Agriculture where about 7500 farmers from central Luzon participated; and a strike that attracted ten thousand protesters at the proposed nuclear power station in the province of Bataan (Schock 2005). The Bataan nuclear power plant was the Philippines' only attempt at building a nuclear power plant. Construction started in 1976 in an earthquake zone, but it was stopped following the 1979 Three Mile Island accident and a subsequent inquiry regarding its safety. The project incurred US$2.3 billion of debt, the country's biggest single obligation and is among the many white elephant projects of the Marcos regime.

In 1986, less than a year after BAYAN was formed, the alliance boasted a national membership of two million, including six hundred thousand *Kilusang Mayo Uno* (KMU; translation: First of May

Movement) members and a hundred thousand members from *Kilusang Magbubukid ng Pilipinas* (KMP, translation: Peasant Movement of the Philippines), and sub-alliances with over five hundred grassroots groups (Schock 2005; Zunes 1999). These organizational and tactical networks across different sectors of Philippine society, in turn, "encouraged quite conservative opposition groups to adopt more radical forms, and provided broad institutional support for anti-dictatorship struggle" (Boudreau 2004: 156). The intense networked mass mobilization and nonviolent resistance grew quickly to involve nearly every segment of Philippine society, including moderate reformers, businesspeople, religious leaders, and even former Marcos regime supporters. The network approach eventually paved the way for greater public engagement that was highly visible during the largely nonviolent 1986 People Power Revolution in Manila.

Network-based mobilizations also became the defining strategy in the US Civil Rights Movement (Darling 2009). In general, scholars argue that the successful change of the racial order in the United States was about recognizing the ability of small groups of black communities to network among themselves (Darling 2009; Du Bois 1996 [1899]). The Movement, one scholar contends, is "more fully realised when viewed as a constellation of localised social movements" (Davis 2001: 4). In the course of civil rights activism, both community leaders and residents mobilized thousands of their own social groups, neighbors, friends, co-workers, and families.

The collective African-American identity in the United States' south can be best viewed mostly within the social networks found in African-American churches, mutual aid societies, benevolent associations, literacy and fellowship societies, and within secret societies (Darling 2009). These collective community institutions, American sociologist and civil rights activist W. E. B. Du Bois (1996 [1899]) argued half a century earlier, was "the sole hope for minority African-Americans in a largely white power-system." In the wake of the Montgomery bus boycott, Du Bois's argument rung true.

In communities, local social groups were organized and linked with the growing Movement. These linkages resulted into events that shook the United States, beginning with Parks' arrest in 1965 and followed by the Montgomery bus boycott and later campaigns such as *Freedom Summer* and the *March on Washington*. These campaigns "were engaged in by relatively powerless groups; and they depended for success not upon direct utilization of power, but upon activating other groups to enter the political arena" (Lipsky 1965: 1). With strong, diverse and large networks, the Movement became self-propelling, especially as the struggle expanded across the United States paving the way to equal rights legislation in 1964.

The history of the Burmese Pro-democracy Movement sat in contrast with these histories. During the critical month of August 1988 when protests already intensified and expanded beyond Rangoon and when the junta's legitimacy appeared to be on the verge of collapse, the Movement was unsuccessful in linking the many heterogeneous groups across the country and in providing an overarching avenue for them to be webbed. Instead, prominent opposition politicians, including Aung San Suu Kyi, elected to operate independently (Zin 2010). The opposition leadership was also internally divided on whether to call for a regime change or support a negotiated transition (Zin 2010). Opposition politicians were also reluctant to form alliances with the original instigators of the uprising: the students and other grassroots opposition groups (Schock 2005).

Recognizing the importance of stronger networks, the students attempted to unite opposition leaders in a single leadership council during the national strike of 26 August 1988; but this too failed (Zin 2010). Ties between the elite political leadership and grassroots groups, instead of becoming strong and united, eventually weakened (Zin 2010; Schock 2005; Boudreau 2004). The failure to form a united national umbrella organization to aggregate and coordinate people's resistance against the junta seriously weakened the Movement during this pivotal stage (Zin 2010; Schock 2005).

The political divisions among Burmese opposition elites were exacerbated when U Nu, one of the main faces of the opposition, attempted to increase leverage by announcing that he was the legitimate prime minister and consequently named a parallel government—a strategy that mimicked Cory Aquino's during the People Power Revolution (Zin 2010). Instead of rallying behind U Nu, opposition leaders, including Aung San Suu Kyi, exhibited growing factions within the political opposition (Zin 2010). With the opposition continuing to show division, a unified regime alternative was not effectively developed for the broader Burmese public. This incident had repercussions in the ranks of grassroots campaigners, and among the people of Burma in general, who became confused, frustrated, and fatigued (Zin 2010; Boudreau 2004). In two years, the struggle failed and the nation returned to strong military control.

Webbing in Today's Climate Action Movement

A networked approach to mobilization is already embraced in the climate action movement (McKibben 2013). Diversity of climate actions are seen across geographic regions, where climate action groups would use multiple campaign approaches. Network-based climate actions have already produced extensive outward-focused climate activism. On 21 September 2014, for example, the power of the networked approach became more evident in one of the largest campaigns launched by the Movement when a number of climate action groups, led by 350.org, coordinated the People's Climate March (Foderaro 2014).

Webbing climate actions, nonetheless, have received mixed sentiments and views from many actors and groups within the Movement. Having a shared notion to potentially create a common ground is regarded as its most important benefit. Webbing climate actions are also seen to enable varied themes to be interconnected and for different groups and actors from various struggles and social contexts to join in one common struggle (Della Porta and Diani 2006).

The concept of multi-stakeholderism has been tagged as necessary to build more inclusive, participatory, and heterogeneous climate actions. For webbing to work, however, plurality, per se, should not be the singular focus, but also on respecting differences in identities, interests, roles, and responsibilities of multiple "stakeholders." The resultant condition brought about by webbing climate actions, otherwise, would be one of the most important caveats to note in mobilizing for climate. In my survey of contemporary social action movements, some of my respondents, for instance, emphasize the "danger" of a divided Movement in what can be called a "we-them" dichotomy (Delina 2018a, Chapter 5; cf. Saunders 2008).

While webbing is ideal as shown by histories of mobilization, power imbalances arising from webbing, indeed, could not be easily ignored. Groups and organizations, normally at odds, may, at some stage, make a "marriage of convenience" to support a particular issue; these alliances, however, can prove disastrous. The failure of the United States Climate Action Partnership (USCAP) to advance an agenda to get a legislated price on carbon in the United States provides an illustration.

USCAP is an alliance of leading environmental movements in the United States and businesses formed to push for a climate legislation that would enable a cap-and-trade system. USCAP included the prominent

Environmental Defense Fund, World Resources Institute, and National Resources Defense Council in its membership roster, and those in the fossil fuel business such as Rio Tinto, Duke Energy, and Shell. In the end, USCAP failed to advance its objective in the US Congress. In her analysis of this failure, Theda Skocpol (2013) argues that if only these highly respected environmental groups had focused on mobilizing grassroots support, instead of allying with and making an "insider deal" with the fossil fuel industry, results could have been different. Skocpol based her assertion by contrasting the efforts made by groups who lobbied for universal health care in America, later known as Obamacare, with that of USCAP. The Congress, at the time, simultaneously heard these two contentious issues.

Skocpol argues that the success of Obamacare was due to a strategy that linked grassroots players with those at the state and national actors through a network that would promote the idea of a public option. The public option was a proposal that would make cheaper health insurance available to uninsured Americans who cannot afford private insurance. It does this by creating a government-run health insurance agency that would compete with private insurers. Although the public option was not included in the final version of the bill, Skocpol suggests that the idea successfully provided broader public support for Obamacare.

Journalist and activist Naomi Klein shares Skocpol's view. In an interview, Klein remarked:

> Their [pertaining to USCAP] so-called win-win strategy has lost. That was the idea behind cap-and-trade. And it was a disastrously losing strategy. The green groups are not nearly as clever as they believe themselves to be. They got played on a spectacular scale. Many of their partners had one foot in [USCAP] and the other in the U.S. Chamber of Commerce. They were hedging their bets. And when it looked like they could get away with no legislation, they dumped [USCAP] completely. (Naomi Klein interview, see Mark 2013; also see Klein 2014: 226–228)

Webbing Tomorrow's Climate Actions

The climate action movement needs to mediate and bridge the current gaps between effective climate actions and the myopic, less stellar climate response by governments and markets. The key challenge for the Movement is not that it does not have the numbers; it is about organization. The Movement has already shown the world it can organize popular mobilization. Climate action networks are, indeed, already far-reaching, with many being maintained on a continuous basis among climate action groups.

The Dakota Access Pipeline protest, the international 2014 People's Climate March, and community energy networks are a few examples among many. The challenge remains, however, to see these networks and their mobilizations continually deployed and strategically sustained to match the power and resources of the fossil fuel regime complex. The time is ripe for the Movement to move beyond horizontal and local-based organizing approaches toward the construction of an expansive, large-scale climate action project that would counter the key driver of anthropogenic climate change: the neoliberal hegemony.

In constructing this counter-hegemony, network-organized, outward-focused climate actions, which are increasingly becoming common, have to be complemented by webs of networks that are focused on more prefigurative activism, i.e. climate actions that prefigure alternative futures. David Hess (2018), for example, suggests building strong energy transition multi-coalitions. There already exist models for achieving this end, including examples from community energy groups that are collectively working and coordinating their activities (e.g. Bristol Energy Network and the Community Energy Scotland Network in the UK [also see Parag et al. 2013; Parag and Janda 2014]), various energy networks in The Netherlands (Van der Schoor et al. 2016), and the 100% Nachhaltige Energie Regionen in Germany (Beveridge and Kern, 2013). The key in this strategy is to link the fragmented pockets of climate actions to achieve scaled impact. Multiple co-benefits accrue from webbed prefigurations, which include: (1) growing stock of learned good practices of climate actions; (2) knowledge exchange of practices to improve local sociotechnical expertise; (3) rollouts of funding for parallel projects; (4) building reflexivity in experimentation, including rewarding leadership through recognition and benchmarking; (5) magnified political pressure to raise ambitions at other levels; (6) deepening cooperation; and (7) promoting "outgrowth" and diffusion (Kern and Bulkeley 2009; Hakelberg 2014).

Scaling up climate actions, however, cannot only be conceptualized around the narratives of replication, that is, local-based innovations mimicked and then transferred in other spaces and contexts, places and localities. The idea that climate actions can *travel* spaces and timelines across communities is a weak proposition especially in terms of its possible failure to acknowledge context-specificities of these actions. The Movement, thus, needs to think beyond replication and, instead, consider webbing heterogeneous and varied, yet healthy and sustaining, ecosystems of multiple, polycentric organizations. Organizing these messy and contested webs is essential since climate actions as sociopolitical-economic-technical projects require not only series and parallel activities but also division of labor.

Simultaneous climate actions can range from increasing the quantity and strengthening the quality of public information through innovative awareness-raising projects to policy and project proposals to legal support to media relations to social media and face-to-face campaigning to prefigurative actions on the ground to name a few. No single group is sufficient to perform all these roles. As demonstrated by histories under study, no single organization carried the mobilization; they were results of broad and wide ecologies of organizations, participation and engagement.

The horizontal architecture of the climate action movement entails that no one person, group, organization or institution should be dominating the Movement. This is true although "leadership occurs as an event in those situations in which some initiatives manage to momentarily focus and structure collective action around a goal, a place or a kind of action" (Nunes 2014: 35). Histories of mobilizations, which produced prominent voices the likes of Gandhi, Martin Luther King Jr., Cory Aquino, and Aung San Suu Kyi, are illustrative of charismatic movement spokespeople; there was, however, no single, leading individual, or group. The climate action movement requires the same "ecology of organization," where leadership is distributed rather than concentrated; meaning that mobilizations can arise from anywhere, where pluralism, cosmopolitanism, heterogeneity, and variety become its strength (cf. Ross 2011).

This kind of diversity within the Movement does not, at all, mean the absence of a coherent vision of a future that everyone can imagine, support, and work around. Indeed, it is the adherence to a coherent vision that gives any Movement the consistency and unity it direly needs to succeed. In the same vein, the climate action movement would deliberately require a "common, overarching, fundamental" vision of what that future *ought* to look like between and within differences and particularisms in the nodes and pockets of popular participation (see Chapter 3), rather than focusing on leadership and loose, uncoordinated campaigns (Carroll 2006). The Movement needs to pitch this organizing idea "not as exclusive but as complementary, whose effects can reinforce each other" (Nunes 2014: 30). The overarching architecture of the Movement, thus, is a relatively webbed, polycentric form (Ostrom 2010; cf. Jordan et al. 2015).

Polycentricity in the climate action movement is ideal in that this arrangement moves decision-making closer to on-the-ground actors who are doing the campaign, prefigurative climate actions, and those who are impacted by climate change. Benefits of this arrangement accrue in terms of equity, inclusivity, information, accountability, organizational multiplicity, and adaptability (Sovacool 2011). The transnational aspect of the Movement also reflects well with the ideals of polycentrism.

To be clear and to borrow Nobel laureate Ellinor Ostrom's (2010) words, there is no perfect governance arrangement. The Movement, thus, could not claim its polycentricity as a privileged institutional arrangement. Indeed, the Movement has to be seen squarely as a consistent hegemonic project aimed at building a solid constituency from multiple and fragmented publics that will work together in advancing a counter-hegemony to neoliberal capitalism. This way, the webbed climate action movement could critique the almost unquestioned 150-year carbon-intensive ideology of progress. Here, the industrial economy built on the rapid exploitation of fossil energy (Malm 2016), and which led to unfairness, inequality and maldistribution of wealth (Piketty 2014) is duly and aptly questioned.

The new social order, advanced by this counter-hegemony, is, thus, emancipatory in nature, constructively transforming societies and fusing multiple "common good" interests with that of a safer climate end. This web of interests will include: justice, equality, democracy, peasants' rights, struggles against patriarchy, defense of indigenous people's rights, civil rights, sustainability, fair trade, food sovereignty, etc. In other words, the climate action movement becomes a Movement of all Movements.

Webbed within this imaginary of different, yet durable, emancipatory and sustainable futures, the Movement should work not on divisions and quarrels between short- and long-term desires, spontaneous- and planned-campaigns, and differing narratives and stories across the heterogeneous groups and organizations that advance them. The Movement, instead, should forcefully address the justification of the many ongoing futile market reform advocacies established to extend the life of fossil fuel-based capitalism. These advocacies include technologies, mechanisms, and approaches that violate the precautionary principle (e.g. geoengineering and nuclear energy), create land-grab postures (e.g. biofuels and even renewable energy systems such as wind, water and sunlight energy built over environmentally and culturally protected areas and fertile agricultural lands), and excessive capitalization cost (e.g. large hydropower dams, nuclear, coal with carbon capture and storage, and fracking), among others. The Movement, indeed, has to be seen as a highly dynamic, yet reflexive, institution acting and reacting, while making decisions in their context-specific challenges. Diversification notwithstanding, climate actions must always cohere with the Movement's emancipatory objectives for structural change that would lead to equitable, just, and sustainable future for all.

In closing, the Movement should afford a wide range of different organizational and tactical compositions, yet they must be webbed together under one coherent vision to provide an overarching imaginary. This means that many climate action groups may burst into spontaneous, yet varied, protests to advance this new hegemony. They will continue to make intricate climate action-related demands (outward-oriented climate actions) on one hand, and organize activities and create new processes to demonstrate what a future could look like such as in community energy, direct democracy, deliberative democracy and other prefigurative climate actions (Delina 2018b).

The Movement not only takes all these activities into account but also adopts a strategic perspective to transform spectacular scenes of protests (as in the Dakota Pipeline protests) and broad community energy (as demonstrated in Germany's and Denmark's energy cooperatives) into effective, long-term, and sustained climate actions. As a long-term institution, the Movement needs to strategize around these (largely ephemeral) multiple tactics and campaigns. The four histories covered in this book demonstrate how street demonstrations are supposed to be complemented by a wide array of organizing and institutionalizing long-term visions, while giving voices to people, creating new identities, and generating clear alternatives.

References

Andrews, K. T., Ganz, M., Baggetta, M., Han, H., & Lim, C. (2010). Leadership, membership, and voice: Civic association that work. *American Journal of Sociology, 115,* 1191–1242.

Beveridge, R., & Kern, K. (2013). Energiewende in Germany: Background, developments and future challenges. *Renewable Energy Law Policy Review, 4,* 3.

Boudreau, V. (2004). *Resisting Dictatorship: Repression and Protest in Southeast Asia.* New York: Cambridge University Press.

Carroll, W. K. (2006). Hegemony, counter-hegemony, anti-hegemony. *Socialist Studies, 2,* 30–32.

Chenoweth, E., & Stephan, M. J. (2011). *Why Civil Resistance Works: The Strategic Logic of Nonviolent Conflict.* New York: Columbia University Press.

Darling, M. (2009). Early pioneers. In M. Ezra & P. Macall (Eds.), *Civil Rights Movement: People and Perspectives.* Santa Barbara: ABC-CLIO.

Davis, J. (Ed.). (2001). *The Civil Rights Movement.* Malden, MA: Blackwell Publishers.

Delina, L. (2018a). Webbing. In *Climate Actions: Transformative Mechanisms for Social Mobilisation.* Cham: Palgrave Macmillan.

Delina, L. (2018b). Climate mobilizations and democracy: The promise of scaling community energy transitions in a deliberative system. *Journal of Environmental Policy & Planning.* https://doi.org/10.1080/15239 08x.2018.1525287.

Della Porta, D., & Diani, M. (2006). *Social Movements: An Introduction* (2nd ed.). Malden, MA, USA: Blackwell.

Du Bois, W. E. B. (1996 [1899]). *The Philadelphia Negro: A Study.* Philadelphia: University of Pennsylvania Press.

Foderaro, L. W. (2014, September 21). Taking a call for climate change to the streets. *The New York Times.*

Gandhi, M. K. (1930). *Collected Works of Mahatma Gandhi* (Vol. 43). New Delhi: Digital Library of India.

Hakelberg, L. (2014). Governance by diffusion: Transnational municipal networks and the spread of local climate strategies in Europe. *Global Environmental Politics, 14,* 107–129.

Hess, D. (2018). Energy democracy and social movements: A multi-coalition perspective on the politics of sustainability transitions. *Energy Research & Social Science, 40,* 177–189.

Jordan, A. J., Huitema, D., Hilden, M., van Asselt, H., Rayner, T. J., Schoenefeld, J. J., et al. (2015). Emergence of polycentric climate governance and its future prospects. *Nature Climate Change, 5,* 977–982.

Kern, K., & Bulkeley, H. (2009). Cities, Europeanization and multi-level governance: Governing climate change through transnational municipal networks. *Journal of Common Market Studies, 47,* 309–332.

Klein, N. (2014). *This Changes Everything: Capitalism vs. the Climate.* New York: Simon & Schuster.

Komisar, L. (1987). *Corazon Aquino: The Story of a Revolution.* New York: George Braziller.

Lipsky, M. (1965). *Protest and City Politics.* Chicago: Rand McNally & Co.

Malm, A. (2016). *Fossil Capital: The Rise of Steam Power and the Roots of Global Warming.* London: Verso.

Mark, J. (2013). Conversation: Naomi Klein. *Earth Island Journal, 28*(Autumn), 45–47.

McKibben, B. (2013, January 14). Beyond baby steps: Analysing the cap-and-trade flop. *Grist*.

Murphy, G. (2005). Coalitions and the development of the global environmental movement: A double-edged sword. *Mobilization, 10*, 235–250.

Nunes, R. (2014). *Organisation of the Organisationless: Collective Action After Networks*. London: Mute.

Ostrom, E. (2010). Polycentric systems for coping with collective action and global environmental change. *Global Environmental Change, 20*, 550–557.

Parag, Y., Hamilton, J., White, V., & Hogan, B. (2013). Network approach for local and community governance of energy: The case of Oxfordshire. *Energy Policy, 62*, 1064–1077.

Parag, Y., & Janda, K. B. (2014). More than filler: Middle actors and socio-technical change in the energy system from the "middle-out". *Energy Research & Social Science, 3*, 102–112.

Piketty, T. (2014). *Capital in the Twenty-First. Century*. Cambridge, MA: Belknap Press.

Rhodes, R. A. W. (1996). The new governance: Governing without government. *Political Studies, 44*(4), 652–667.

Ross, C. (2011). *The Leaderless Revolution: How Ordinary People Will Take Power and Change Politics in the Twenty-First Century*. New York: Blue Rider Press.

Saunders, C. (2008). Double-edged swords? Collective identity and solidarity in the environmental movement. *The British Journal of Sociology, 59*, 227–253.

Schock, K. (2005). *Unarmed Insurrections: People Power Movement in Nondemocracies*. Minneapolis: University of Minnesota Press.

Skocpol, T. (2013, February 14). Naming the Problem. In *Harvard University Symposium on the Politics of America's Fight Against Global Warming*. Harvard University.

Sovacool, B. K. (2011). An international comparison of four polycentric approaches to climate and energy governance. *Energy Policy, 39*, 3832–3844.

Van der Schoor, T., van Lente, H., Scholtens, B., & Peine, A. (2016). Challenging obduracy: How local communities transform the energy system. *Energy Research & Social Science, 13*, 94–105.

Van Dyke, N., & McCammon, H. J. (2010). *Strategic Alliances: Coalition Building and Social Movements*. St. Paul: University of Minnesota Press.

Zin, M. (2010). Opposition movements in Burma the question of relevancy. In S. L. Levenstein (Ed.), *Finding Dollars, Sense, and Legitimacy in Burma*. Washington, DC: Woodrow Wilson International Center for Scholars.

Zunes, S. (1999). The origins of people power in the Philippines. In S. Zunes, S. B. Asher, & L. L. Kurtz (Eds.), *Non-violent Social Movements: A Geographical Perspective* (pp. 129–157). Malden: Blackwell.

CHAPTER 8

Conclusion: Strengthening Climate Actions Through Emancipatory and Transformative Mobilizations

Abstract The contemporary climate action movement can be strengthened by reorienting its campaigns toward becoming hegemony-changing exercises. To achieve this grand vision, histories of past social mobilizations suggest key mechanisms that the Movement can add in its campaign repertoires and initiate new cultures of progressive climate activism. These mechanisms are: visioning and identity-building; culturing and framing; peer pressure; diversity; publicity; and networks. Using these mechanisms, the Movement can mobilize its constituency through stronger forms of traditional protests while expanding new opportunities for prefiguring desirable futures across multiple and heterogeneous spaces and places. All climate actions, nonetheless, need to be accomplished with an end-view of achieving the grand vision of replacing neoliberal capitalism with a new hegemony based on principles of just emancipation and sustainable transformations.

Keywords Mobilization · Hegemony · Neoliberalism · Climate action · Just emancipation · Sustainable transformation · Heterogeneity · Justice · Climate action movement

Time after time and place after place, publics mobilize and organize into Movements to emancipate themselves and carry on the tradition of transformations. In the seaside town of Dandi, in the saltworks of Dharasana, in the villages of Gujarat, in the neighborhoods of Montgomery, in the

© The Author(s) 2019
L. L. Delina, *Emancipatory Climate Actions,*
https://doi.org/10.1007/978-3-030-17372-2_8

111

streets of Manila, and in the colleges of Rangoon, publics demonstrated that the rule by unjust regimes were forms of domination and that public consent on their power could be revoked. Similar grassroots mobilizations, operating under strict nonviolent discipline, can be scaled for transformative and emancipatory Movements. Indeed, this book has shown the historical sophistication, ability, and capacity of people-driven and nonviolent social action campaigns to successfully crack elite and powerful, yet unjust, regimes.

In the wake of the many failures, myopia, and ineffective climate actions, nationally and internationally, this book offers the mirrors of the past to explore the opportunities for strengthening and establishing a greater momentum for the grassroots component of future climate actions so that they can become more effective, and eventually succeed and prosper. This book recommends some strategies for mobilizing publics into climate action groups and organizations, and bringing them together into webs of stronger sociopolitical forces that is an institutionalized climate action movement. Historical analysis of four select histories—from India to the United States to the Philippines to Burma—showed that campaign effectiveness is a function of at least five broad mechanisms:

- Visioning and building a new collective identity by suggesting the imperative of a counter-hegemony that is largely based on principles of justice and sustainability;
- Messaging the morality of social actions in the present, and the culturing and framing of the message of these actions by choosing appropriate symbols;
- Triggering communal peer pressure through prefigurative, yet heterogeneous, social actions;
- Encouraging diversity of participants and their campaigns;
- Boosting the publicity of social actions using multiple social communication channels; and
- Diversifying networks and webbing pockets of social actions to achieve scale.

Notwithstanding the events that followed after the specific campaigns in the four histories, this book describes *what* broadly determines and constitutes effective campaigns and *how* these Movements had metamorphosed from pockets of small, locally based resistance into effective, networked, and large-scale mass actions that transformed the way their respective societies were emancipated. The cracks that these episodes left in their respective histories show that regime changes are always possible, regardless of how impenetrable, strong, and solid they might appear. This impression, in many ways, mirrors the strength of the current fossil fuel regime complex and their supporters in governments and mass media—a regime duly constructed within the powerful frames of a neoliberal, capitalist, and expansive order. The cases of Movements are, indeed, illustrative of the almost impossibility of defeating powerful and impregnable regimes; yet, the power of sustained and strengthened nonviolent actions brought them down.

Drawing strategies from historical analysis of these moments for their contemporary use in the climate action movement, however, includes a number of caveats. First, historical and contemporary analogues are never exact. While similarities can be identified, they never are universal. Just like any other Movements, the climate action movement has messy, complex, distinct and unique features. The heterogeneity of climate action groups means that strategies and approaches to climate action could never be exclusive. Second, all social movements are uncertain, unpredictable, and contentious social dynamics; hence, there will never be a predetermined pathway for mobilization and organization on climate actions. Although the strategies advanced in the book for use in climate action mobilizations were designed as aligned with the lessons provided by histories, it is also possible that climate action groups digress and take extraordinary pathways in designing and prosecuting their own and respective campaigns. Uncertainty is a fact and therefore opportunities for new learning, including that of being reflexive and appreciative of messiness and complexity, should be embraced within the Movement.

Despite the caveats presented by the parallel analysis made in this work, there remains highly relevant lessons when bridging the past and the present, especially considering that mobilizations have a long political pedigree to change societies, transform social orders, and emancipate peoples. Therefore, it remains salient, paramount, and relevant that the climate action movement continue their work, regardless of the hardships and moments of failures, to advance a counter-hegemony to neoliberal capitalism: a hegemony that ought to be framed based on the ideals of just emancipations and sustainable transformations for all.

The central argument of this book is about questioning the extreme complex of fossil fuel power and our present society's heavy reliance upon capitalism's self-correction mechanisms. In this new hegemony, justice and sustainability are made more central than peripheral; and sustainable development is not only framed as about the present and future generations of humans and non-humans but also having the elements of prudence, intergenerational equity, precaution, responsibility, and good governance. This new hegemony is not only told in street protests, it is also prefigured in spaces and places where durable and desirable futures are demonstrated and lived by people in the present, not only as examples but more of experimental social laboratories. Here, people and their communities extensively reflect upon their mistakes and failings, at the same time that they celebrate their successes.

That said, this book, to be clear, is not only about a grand vision; it also underlines the need for highly situated and contextual climate actions—which, when well-communicated and webbed, could contribute to meeting the ends of that grand vision. These climate actions are focused on the actual, real-life, social-political-economic conditions of communities, spaces, and geographies that will deliberate and advance those actions. In these inclusive, authentic, and influential deliberations, participants pay heed to—and respect—each other's expertise, experiences, and positions. Facts, including scientific data, are deeply considered rather than mere opinions. People then make decisions based on these facts, and as a collective, would commit to these decisions. These processes of knowing are translated into processes of doing.

This book acknowledges the power of webbed networks. This entails the enrollment and mobilization of allies. At the same time, such enrollment processes recognize that not everything can be webbed. In other words, the context specificities of different climate actions arising from various spaces and places mean that scaling, transfer, and replication are dynamics that are never guaranteed. While full-scale transfers are uncertain, local innovations still carry with them messages and lessons that others in these webs can look at and learn from. A cosmopolitan and networked approach to mobilization, thus, highlights the multiple possible interconnections of the many circuits of practice, thought collectives, regimes of truth, relations and processes, and systems of meanings that are present in the diversity of climate actions.

Climate actions, in closing and to stress, are about revoking consent. The histories of mobilization this book has used are rife with moments where ordinary people, acting spontaneously and using nonviolent methods, revoked their consent to unjust powers. In the same way, contemporary climate actions must be viewed as processes to revoke consent from unjust powers in the fossil fuel complex and the neoliberal hegemony that sustains their power. These climate actions should not only be justly emancipatory but also sustainably transformative.

INDEX

0-9

1.5 °C, 3, 4, 76
100% energy transition, 55
100% Nachhaltige Energie
    Regionen, 104
100% renewable energy, 42, 43
350.org, 102
8888, 8 August 1988, 30, 31

A

accountability, 5, 106
adaptability, 106
agency, 55
ahimsa, 73
Ahmedabad, India, 19, 73
Alabama, 18, 21–23
Alaska, 4
alignment, 78
Alinsky, Saul, 7
alliances, 5, 27, 98, 100, 101
allies, 73, 114
alternative, 12, 28, 35, 37, 41,
    45, 60, 61, 63, 88, 101,
    104, 107

American College & University
    Presidents' Climate
    Commitment, 91
anti-Apartheid movement, 8
anti-racist, 80
Aquino, Cory, ix, 27, 40, 41, 57, 62,
    101, 105
Aquino, Ninoy, ix, 24, 26, 27, 57, 60,
    62, 86, 99
Arctic, 4
artificial intelligence, 62
Asia, 6
Australia, 6, 7, 87, 88
Australian Centre for Independent
    Journalism, 87
autarky, 42
authenticity, 90, 91, 114
automation, 45
autonomous mobility, 45, 62
autonomy, 27, 42, 45, 62

B

backfire, 59, 85
Bagong Alyansang Makabayan, 27, 99

Baroda, India, 38
Bataan nuclear power plant, 99
behavior, 10, 36, 54, 55, 78, 79
belonging, 36, 74
benchmarking, 104
Bengal, 19
big data, 45, 62
biomimicry, 45
Birmingham, Alabama, 23
#BlackLivesMatter, 80
Bloody Friday massacre, 29
Bohol, Philippines, 6
Bombay, India (now Mumbai), 19,
    38, 84
boycott, ix, 17, 19, 21–23, 39, 40, 58,
    74, 75, 85, 99, 100
Bristol Energy Network, 104
British Empire, 18–21, 37, 38, 41,
    73, 99
Brown v. Board of Education, 21, 74
Buddha, 57
Burma, now Myanmar, 18, 29–31, 63,
    101, 112
Burma Socialist Programme Party
    (BSPP), 29

C
Calcutta, India (now Kolkata), 38
California, 5, 59, 76
California wildfires, 59, 76
cap-and-trade, 102, 103
capitalism, 2–4, 8, 10, 12, 25, 44–48,
    62, 63, 106, 113
carbon, 6, 46, 60, 77, 106, 107
carbon capture and storage, 107
carbon price, 4
Catholic Church, 18, 25–27, 40, 86,
    99
Central America, 6
charisma, 22, 105
China, 3
circular economy, 55

civil disobedience, 19, 23, 28, 32, 37,
    38, 40, 99
civil rights, ix, 18, 21, 23, 39, 56, 63,
    74, 85, 90, 100, 106
Civil Rights Act of 1964, 23
Civil Rights Movement, 18, 21, 23,
    63, 100, 105
clicktivism, 90
climate action movement, 5, 7–9, 11,
    12, 36, 41–47, 53, 54, 58–64,
    72, 76, 77, 79, 80, 87–91, 102,
    103, 105–107, 112, 113
climate change, x, 2–4, 6, 7, 9, 10, 12,
    41, 47, 48, 54, 55, 58, 61–63,
    76, 77, 84, 87, 88, 91, 104, 106
climate denial, 4, 6, 10, 44, 46, 54,
    59, 61, 76, 77, 84, 87
Climate Leadership Network, 91
coal, 3, 4, 6, 41, 62, 77, 80, 87, 88, 107
coalitions, 5, 31, 80, 98, 104
cockroach ideas, 54
Colvin, Claudette, 74
common good, 106
Commonwealth Scientific and
    Industrial Research Organisation
    (CSIRO), 7
Communist Party of the Philippines
    (CPP), 25, 40
Community Energy Scotland
    Network, 104
community renewable energy, x, 6, 42,
    43, 61, 62, 72, 91, 104, 107
complacency, 4
conflict, ix, x, 3, 6
consensus, 91
consent, 4, 38, 40, 43, 112, 114
conservative think tanks, 87
consumption, 3, 4, 55, 78
cooperation, 27, 32, 37, 104
cooperatives, 42, 61, 107
cosmopolitanism, 11, 105, 114
The Courier Mail, 88
credibility, 61, 62, 90

cronyism, 24–26
cryptocurrency, 45
culture, 2, 4, 5, 11, 12, 44, 55, 58–62, 79
culturing, 10, 55–60, 63, 112

D

The Daily Telegraph, 87, 88
Dakota Access Pipeline, 104, 107
Dandi, India, ix, 17–21, 37, 39, 72–74, 84
Dandi March, 17, 18, 20, 37, 72–74, 84
David, Randolf, 57
deliberative democracy, 11, 64, 91, 107, 114
democracy, 4, 26, 29–31, 42, 44, 46, 47, 61–64, 101, 106, 107
Democracy Now!, 88, 89
Denmark, 107
despair, 4
Dharasana campaign, ix, 17–21, 39, 85, 111
diffusion, 77, 104
dissonance, 10, 55
diversity, 9, 11, 25, 77, 97, 98, 101, 102, 105, 112, 114
divestment, 6, 7, 41, 62, 77
doomsday, 58
Du Bois, W.E.B., 100
Duke Energy, 103
Duterte, Rodrigo, 4
dystopia, 2, 43

E

ecology of organization, 105
economic literacy, 45
efficiency standards, 5, 6
electricity, 3, 43, 47, 77, 78
emancipation, 5, 44, 63, 64, 90, 91, 106, 107, 111–113

empowerment, 61
endowments, 6, 77
Energiewende, 6, 42
energy efficiency, 6, 42, 61, 78, 88
energy-intensive agriculture, 55
energy transition, 42, 45, 55, 77, 87, 104
Enrile, Juan Ponce, 28
Environmental Defense Fund, 103
Environmental Protection Agency, 5
Epifanio de los Santos Avenue (EDSA), 26, 28, 75, 86
equality, 21, 25, 44, 106
equity, 44, 106, 114
expansionist, 6, 10, 11, 18, 19, 27, 29, 37, 38, 46, 62–64, 73, 88, 97, 101, 104, 113
expectations, 72, 73
experimentation, 45, 104, 114
expertise, 104, 114
extra-judicial killings, 26
extreme weather events, 2, 3, 54, 58–60, 62, 76

F

face-to-face, 79, 84, 90, 99, 105
failure, x, 5, 7–10, 12, 18, 20, 31, 32, 48, 55, 76, 88, 99, 101, 103, 105, 112, 114
fairness, 44, 64, 84, 106
fake news, 11, 90
false balance, 84
famine, 58
fear, 4, 54, 59, 74
Finnmark, 91
Fischer, Louis, 21
folk politics, 11, 43, 45, 62, 63
food, 31, 40, 78, 106
fossil fuel, 1–3, 6, 8, 9, 9, 10, 41–47, 61–63, 77, 79, 80, 84, 87, 88, 103, 104, 106, 113, 115
fossil fuel non-proliferation treaty, 47

Fox News, 87
fracking, 107
framing, 10, 42, 43, 55–59, 61–63,
    76, 78, 79, 87, 112, 113
freedom, 5, 44, 63
Freedom Rides, 23
Freedom Summer, 23, 100

G
Galman, Ronaldo, 24
Gandhi, Mahatma, ix, 17–21, 37–40,
    56, 57, 60, 62, 72–74, 79, 84,
    85, 98, 99, 105
gender, 25, 80, 98
geoengineering, 47, 106
Germany, 6, 42, 104, 107
globalization, 55
governance, 11, 47, 106, 114
Gramsci, Antonio, 44, 92
Green Climate Fund, 5
greenhouse gas emissions, 6, 7, 10,
    60–62, 76, 78, 80
Green New Deal, 46
green politics, 11
green purchasing, 61
Greensboro, North Carolina, 23
Gujarat, India, 19, 38, 73, 111

H
habeas corpus, 25
happiness, 44
hegemony, 2, 11, 12, 44, 45, 47, 62,
    63, 89, 91, 104, 106, 112–115
Hess, David, 104
heterogeneity, 5, 7, 11, 36, 45, 64, 74,
    80, 91, 97, 98, 101, 102, 105,
    106, 112, 113
hope, 43, 58–61, 100
hopelessness, 4
human rights, 21, 24, 26, 27, 56, 62,
    63, 101, 106

I
identity, 10, 11, 36–44, 46, 59–61,
    73, 100, 102, 107, 112
ideology, 2, 11, 92, 98, 106
inclusion, 10, 11, 106
India, ix, 17–21, 37, 38, 41, 47, 56,
    57, 60, 63, 73, 74, 84, 85, 90,
    98, 99, 105, 112
Indian Independence Act of 1947, 18
indigenous peoples, 80, 91, 106
individualism, 55
Indonesia, 3, 6
inequality, 2, 43, 44, 63, 106
in-group information, 79
institutional arrangements, 45
insurance, 77, 103
integrity, 56
interconnections, 11, 45, 98, 114
International Monetary Fund, 25
International Panel on Climate
    Change (IPCC), 3–5, 76, 87
Internet, 41, 46, 88–90

J
Jacobson, Mark, 43
Japan, 3
jobs, 42, 61, 80
Johnson, Lyndon B., 23
Juliana v. United States, 62
justice, ix, x, 12, 21, 43, 44, 46, 56,
    58, 61, 63, 64, 89, 106, 107,
    112, 114

K
Kahan, Dan, 59, 60, 72
Kasturbhai, 20
Keystone XL oil pipeline, 4, 6, 77
Khan, Abdul Ghaffar, 19
Kilusang Mayo Uno, 25, 99
King, Martin Luther Jr., 22, 23, 39,
    86, 105

Klein, Naomi, 103
Krugman, Paul, 54

L
labor union, 25, 27
Lahore, Pakistan, 38
leadership, 11, 74, 101, 104, 105
legislation, 11, 21, 77, 102, 103
lifestyle, 56
lobbying, 77, 98
Lord Irwin, 20
Luzon, Philippines, 99
Lwin, Sein, 30, 32

M
Madras, India, 38
Makati Business Club (MBC), 25–27, 40
Makati, Philippines, 25, 27, 40
Manila International Airport, 24
Manila, Philippines, 24–26, 40, 57, 73, 75, 86, 100, 112
manipulation, 59
March on Washington, 23, 100
Marcos, Ferdinand, ix, 18, 24–28, 40, 41, 57, 75, 86, 99, 100
Marcos, Imelda, 25
martial law, 24–26, 31, 86
Maung, Maung, 30, 31
Maung, Saw, 31
Maw, Phune, 29
McKibben, Bill, 63
media, 3, 5, 6, 11, 24, 28, 41, 44, 46, 56, 62, 83–90, 105, 113
messaging, 10, 54, 55, 58, 61, 62
Midnapur, India, 19, 38
Miller, Webb, 85
Mindanao, Philippines, 99
Mississippi, 23
modeling, 45

Mohammad, 57
Montgomery, Alabama, ix, 18, 21–23, 39, 56, 57, 60, 73–75, 85, 100, 111
morality, 37, 40, 54, 56–59, 62, 63, 85, 112
multi-stakeholderism, 102
Murdoch, Rupert, 87, 88

N
Naidu, Sarojini, 20
nanotechnology, 46
narratives, 10, 59, 61, 63, 77, 79, 92, 104, 106
National Association for the Advancement of Colored People (NAACP), 22, 56, 74, 75
National League for Democracy (NLD), 29
Nationally Determined Contributions, 2
National Movement for Free Elections (NAMFREL), 27
National Resources Defense Council, 103
natural gas, 41
NBC Nightly News, 88
Nehru, Jawaharlal, 20, 37
neoliberalism, 2–4, 8, 10, 12, 44, 45, 47, 48, 63, 104, 106, 113
nepotism, 24
Netherlands, 62, 104
networks, 5, 11, 25, 74, 79, 88, 89, 98–104, 112, 114
new economics, 12, 45, 64
New Economics Foundation (NEF), 45
New People's Army (NPA), 25, 40
News Corporation, 88
New Systems Project, 45
New York Post, 87

Nixon, Edgar Daniel, 56
non-humans, 11
nonviolence, ix, x, 6, 7, 18–23, 27, 28,
    30–32, 39–41, 45, 57, 60–62, 73,
    85, 98–100, 112–114
norms, 5
North Carolina, 23
nuclear energy, 4, 47, 99, 106, 107
nudge, 11, 46, 72, 77, 78

O
Obama, Barack, 4, 62
Obamacare, 103
obligation, 10, 71, 75–77, 79, 99
organic intellectuals, 2, 44, 92
organizing, 7, 10, 40, 41, 104, 105,
    107, 113
outward-oriented protests, 6, 107
ownership, 44, 47, 91

P
Pakistan, 18
Paris Agreement, 2, 4, 6
Parks, Rosa, ix, 18, 21–23, 56, 57, 60,
    74, 75, 86, 100
passive design, 45
Patidars, 38
peer pressure, 10, 71–75, 77–79, 112
pension funds, 6, 8, 41, 77
People Power Revolution, x, 18, 28,
    40, 63, 86, 89, 100, 101
People's Climate March, 88, 102, 104
Peshawar, 19
Philippines, ix, xi, 4, 6, 18, 24–26, 28,
    31, 40, 41, 47, 57, 60, 62, 63,
    75, 86, 89, 90, 99, 100, 112
pluralism, 10, 45, 105
polar bear, 54, 58
police brutality, 29
polycentric governance, 11, 98, 105,
    106

populism, 4
Post-Crash Economic Society, 45
post-work, 45
poverty, 25, 63
power, 2, 4–7, 10, 20, 21, 24, 26,
    28, 31, 37, 40, 42–45, 61, 73,
    77, 78, 80, 86, 99–102, 104,
    112–115
precaution, 114
prefigurative dissent, x, 71, 73, 77, 88,
    104–107, 112, 114
Proclamation 1081, 24
prosperity, 44
proximity, 59
prudence, 114
psychology, 54, 55, 58–60, 72, 76,
    78, 79
public education, 45
public engagement, 7, 9, 37, 64, 76,
    79, 100
public funding, 46, 47
publicity, 11, 83, 85, 86, 88, 112
public opinion, 20, 84, 89
public option, 103
public participation, 19, 27, 31, 36,
    37, 42, 47, 61, 62, 64, 71, 75,
    77, 79, 83, 91, 98, 99, 102, 105,
    112, 114
public perception, 7, 62, 76
public transport, 7, 22, 61, 78

R
Radio Veritas, 26, 28, 86
Radyo Bandido, 86
Rama, 56
Ramos, Fidel, 28
Rangoon, Burma (now Yangon,
    Myanmar), x, 18, 29, 30, 73,
    101, 112
Rangoon Institute of Technology
    (RIT), 29
Rangoon University, 29, 30

reflexivity, 10, 36, 45, 64, 90, 104, 107, 113, 114
regulations, 6, 60
re-nationalization, 47
renewable energy, 1, 6, 7, 10, 11, 42–44, 46, 47, 77, 87, 88, 107
replication, 104, 114
resiliency, 20, 42
responsibility, 54, 114
Rethinking Economics, 45
Rio Tinto, 103
risk perception, 72
Russia, 5

S
Sabarmati Ashram, 19, 73
salt tax laws, 18–20, 72, 74
Sami, 91
satyagraha, 19, 74, 99
Saudi Arabia, 5
scale, 2, 5, 8–11, 20, 21, 28, 30, 32, 42–44, 46, 47, 54, 59, 62, 63, 73, 75, 76, 79, 80, 88, 89, 97, 99, 103, 104, 112–114
Scokpol, Theda, 103
Seattle, Washington, 6
self-denial, 73
Sharp, Gene, 7
Shell, 6, 103
shocking events. *See* tipping point
Shwedagon Pagoda, 30
Sin, Jaime, 26, 28, 57, 86
Skocpol, Theda, 103
Smith, Mary Louise, 74
social capital, 90
social contract, 62
social media, 11, 41, 62, 83, 84, 88, 90, 105
social ties, 72
South Africa, 8, 19

South Cotabato, Philippines, 6
Sri Lanka, 57
Stanford University, 6, 43
state planning, 46
Stirling, Andy, 47, 55, 60
stories, 7, 10, 41–43, 55, 61, 79, 84, 87, 88, 106
suburbia, 55
success, 5, 7–9, 114
Sunrise Movement, 77
superannuation funds, 77
Surat, India, 19
sustainability, ix, x, 2, 12, 41–46, 61, 63, 64, 77, 87, 89, 92, 106, 107, 114
Suu Kyi, Aung San, 31, 101, 105
swaraj, 73
symbols, 10, 19, 54, 56–60, 74, 90, 112

T
Tatmadaw, 30, 31
tax, 6, 24, 60, 78
tax, on carbon, 6
Texas, 6
Thailand, 6, 42
Three Mile Island accident, 99
Time Magazine, 20
tipping points, 4, 55, 63
Trade Union Congress of the Philippines, 25
transformation, 2, 4, 5, 8, 10, 36–39, 41, 44–47, 55, 60, 73, 90, 91, 106, 107, 111–113
transition, ix, x, 2, 30, 42, 44–47, 55, 77, 87, 101
Transition Towns, 42
Trump, Donald, 4, 62
trust, 79
Turkey, 3

Typhoon Haiyan, 59

U
unborn, 11
uncertainty, 113
United Democratic Opposition
  (UNIDO), 27
United Kingdom (UK), 104
United Nations Framework
  Convention on Climate Change,
  3
United States Chamber of Commerce,
  103
United States Climate Action
  Partnership (USCAP), 102, 103
United States (U.S.), 3–6, 21, 23, 24,
  60–62, 74, 77, 85, 87, 88, 90,
  91, 100–103, 112
unity, 32, 79, 101, 105
urban poor, 25, 27, 30
urgency, 5, 76, 79
utopia, 43, 44, 62

V
values, 37, 44, 47, 54, 56, 59, 60, 63,
  64

vision, 2, 4, 10, 35–37, 39–44, 46, 80,
  92, 105, 107, 112, 114
voting rights, 21
Voting Rights Act of 1965, 24

W
Wall Street Journal, 87
webbing, 2, 9–11, 42, 91, 97, 98,
  101–107, 112, 114
Win, Ne, 29, 30
Woolworth department store, 23
work, 4, 9, 23, 31, 43–47, 55, 56, 62,
  72, 87, 98, 106, 113
World Bank, 25
World Resources Institute, 103
World War 2, 46

Y
Yale Program on Climate Change
  Communication, 6

Printed in the United States
By Bookmasters